AF615907

Preparation and Properties of Solid State Materials

VOLUME 7

Growth Mechanisms and Silicon Nitride

PREPARATION AND PROPERTIES OF SOLID STATE MATERIALS

Volume 1 / Aspects of Crystal Growth, *edited by Robert A. Lefever* (out of print)

Volume 2 / Chemical Vapor Transport, Secondary Nucleation, and Mass Transfer in Crystal Growth, *edited by William R. Wilcox*

Volume 3 / III-V Alloys, Convective Instabilities, and Nucleation, *edited by William R. Wilcox and Robert A. Lefever*

Volume 4 / Morphological Stability, Convection, Graphite, and Integrated Optics, *edited by William R. Wilcox*

Volume 5 / Floating-Zone Silicon, *Wolfgang Keller and Alfred Mühlbauer*

Volume 6 / Lead Tin Telluride, Silver Halides, and Czochralski Growth, *edited by William R. Wilcox*

Volume 7 / Growth Mechanisms and Silicon Nitride, *edited by William R. Wilcox*

Additional volumes in preparation

Preparation and Properties of Solid State Materials

VOLUME 7

Growth Mechanisms and Silicon Nitride

Edited by

WILLIAM R. WILCOX

Department of Chemical Engineering
Clarkson College of Technology
Potsdam, New York

MARCEL DEKKER, INC. New York and Basel

Library of Congress Cataloging in Publication Data
Main entry under title:

Growth mechanisms and silicon nitride.

(Preparation and properties of solid state materials; v. 7)
Includes bibliographical references and index.
1. Crystals--Growth. 2. Silicon nitride. I. Wilcox, William R. II. Series.
TA401.P655 vol. 7 [QD921] 620.1'1s 81-17486
ISBN 0-8247-1368-0 [548'.5] AACR2

MARCEL DEKKER, INC.
270 Madison Avenue, New York, New York 10016

Current printing (last digit):
10 9 8 7 6 5 4 3 2 1

PRINTED IN THE UNITED STATES OF AMERICA

PREFACE

This is the concluding volume in this series of advances on the preparation of solids and on the influence of preparative conditions on properties. It deals with two very important topics: growth spirals and their relation to the atomic mechanisms of crystal growth; and preparation, properties and applications of silicon nitride ceramics.

Many thanks to all those who have contributed to this series--the staff at Marcel Dekker, the typists, and especially the authors. I hope our readers have learned as much as I have.

William R. Wilcox

CONTRIBUTORS

P. BENNEMA, RIM Laboratories of Solid State Chemistry, Catholic University of Nijmegen, Nijmegen, The Netherlands

WILLIAM J. CROFT, Army Materials and Mechanics Research Center, Watertown, Massachusetts

DONALD R. MESSIER, Army Materials and Mechanics Research Center, Watertown, Massachusetts

I. SUNAGAWA, Institute of Mineralogy, Petrology, and Economic Geology, Tohoku University, Aoba, Sendai, Japan

CONTENTS

Preparation and Properties of Solid State Materials

VOLUME 7

Growth Mechanisms and Silicon Nitride

Chapter 1

MORPHOLOGY OF GROWTH SPIRALS: THEORETICAL AND EXPERIMENTAL

I. Sunagawa

Tohoku University
Aoba, Sendai
Japan

and

P. Bennema

Catholic University of Nijmegen
Nijmegen
The Netherlands

I. INTRODUCTION

In 1959, Burton, Cabrera, and Frank (hereafter BCF) published their now famous paper on "The Growth of Crystals and the Equilibrium Structure of Their Surfaces" [1]. This paper represented an enormous breakthrough in the field of crystal growth, because for the first time a more or less complete crystal growth theory was developed, which could serve as a basis to interpret observations of growth kinetics and step patterns on crystal surfaces.

As soon as Frank put forward the spiral growth theory [2], Griffin [3] presented the first direct observation of a growth spiral on the prism face of a natural beryl crystal. The observation was followed by numerous reports on observations of growth spirals by many authors, using various techniques, such as reflection and transmission microscopy, phase contrast microscopy, two-beam and multiple-beam interferometry, and replication techniques of electron microscopy. Earlier observations and interpretations were summarized in the books of Verma [4] and Dekeyser and Amelinckx [5], in which many beautiful examples are illustrated, including SiC, ZnS, CdI_2, mica, apatite, and paraffin. Both books also dealt with spiral morphology, interaction between neighboring spirals, effect of movement of screw dislocations during growth upon spiral morphology, polytypism and interlacing patterns, etc.

The principal techniques used during this period were phase contrast microscopy and multiple-beam interferometry, which have high resolution power in the vertical direction [4-6]. Growth spirals having the height of 1 unit cell, its small multiples, or even fractions thereof were observed on a wide variety of crystals, definitely providing convincing evidence of the validity of Frank's mechanism. The smallest height of spiral layers so far reported is 2.3 Å, corresponding to a layer with the smallest possible height in the structure of Fe_2O_3, which was observed and measured by Sunagawa [7] on the (0001) faces of natural hematite Fe_2O_3.

Although these optical techniques have a high resolution power in the vertical direction, the resolution is seriously limited in the lateral direction. In the 1960s, Bethge and his group developed and applied the gold-decoration technique of electron microscopy, which enabled them to increase the resolution power both in horizontal and vertical directions.[8-10]. They mainly studied evaporation processes on NaCl samples and showed beautiful step patterns, both spiral and two-dimensional types. This technique was applied by several workers to tiny crystals of natural minerals such as clay minerals [11,12]. Kaishev, Budevski, and coworkers have also made beautiful observations of spirals during electrolytic growth of silver crystals [13].

In the Burton-Cabrera-Frank (BCF) theory, a rather complicated nonlinear differential equation is derived. No exact solution of this equation is possible, but good approximations and numerical solutions for the spiral can be obtained. Beginning about 1970, computer simulation studies have been initiated in the field of crystal growth to simulate the atomistic processes of crystal growth. Spiral growth has also been introduced in such simulations (see references in Sec. II). Thanks to these computer simulation studies, the spiral theories have become more mature. Since we now also have ample observations on spiral morphology, it seems an appropriate time to let the results of the computer simulations confront the observations. It is therefore the purpose of this chapter to select appropriate examples of spiral morphology and to compare them with the theoretical morphology of growth or dissolution spirals.

For this purpose, we shall at first summarize the results of the spiral theory of BCF, together with the results of the recent new developments of the spiral theory, including Monte Carlo simulations of spiral growth. Next, the results of these theories will be compared with a variety of spirals observed by one of us (I.S.) on a variety of both natural and synthetic crystals. It will be shown that such phenomena as polygonization of spirals away from the center and in the center, spirals having rough steps, etc. can be explained in principle on the basis of modern spiral theories.

The presentation is divided into two parts: a theoretical part (Sec. II) and an experimental part (Sec. III). In the theoretical part first a brief survey is given of recent statistical models of surfaces and especially steps. The attention is focused on the roughness of steps. Then the spiral theory and its recent developments are reviewed briefly.

In the experimental part, first a brief explanation is given of the sophisticated methods of observation and measurement used to detect growth spirals of monomolecular heights.† Then actual examples of various shapes of growth spirals are shown and explained following the order of explanation in the theories reviewed in Sec. II.

Before starting our discussion, it is perhaps useful to discuss the supersaturation regions under which spiral growth occurs and their relation with other growth mechanisms (such as two-dimensional nucleation) and unstable growth (hopper and dendritic growth).

Judging from the supersaturation versus growth rate relation, there is a transitional (critical) supersaturation σ^* below which the spiral growth mechanism operates and above which two-dimensional nucleation is dominant (Fig. 1). At still higher supersaturations, a second transitional supersaturation σ^{**} is expected above which unstable growth occurs, which corresponds to dendritic growth [14,15]. The value of σ^* depends on the materials, growth conditions, and type of phases from which crystals grow. It is assumed that σ^* is much higher (25-50% [1]) for vapor growth than for solution growth (for which σ^* is on the order of a few percent or less [14]), which suggests that there is a greater opportunity to meet growth spirals on the surface of vapor-grown crystals than solution-grown crystals. This has in fact been demonstrated experimentally [15]. Both natural and synthetic crystals grown from vapor phases almost invariably show growth spirals, whereas those from solution phases show spirals less often.

---

†The term *monomolecular* will be used in a broad sense. Monomolecular layers mean growth layers having a step height either monoatomic or 1 unit cell.

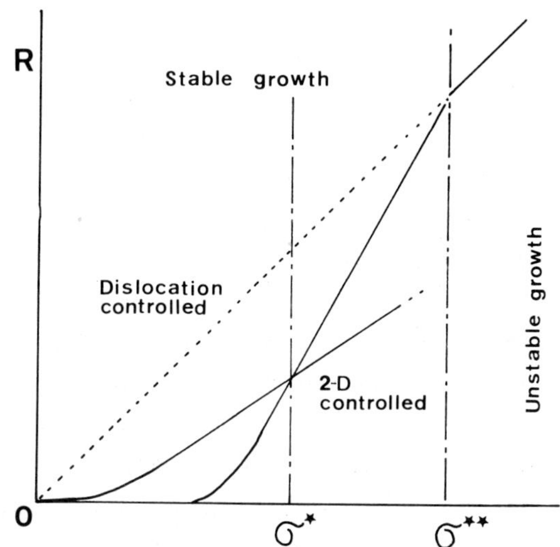

FIG. 1. Schematic presentation of growth rate R vs. supersaturation σ relation. Below σ^* growth is controlled by dislocation, between σ^* and σ^{**} by two-dimensional nucleation; above σ^{**} unstable growth (dendritic growth) takes place.

The theories and observations to be presented in this chapter are therefore directly related to crystals grown under supersaturations lower than σ^*.

II. THEORY

A. Survey of Statistical Models of Surfaces and Steps

1. *Lattice-Gas Model; Step Model*

A frequently used model for the crystal-vapor interface system is the lattice-gas model. The system is divided into an imaginary lattice of cells, each cell being either filled or empty. When attractive forces are present between filled cells, phase separation occurs below a critical temperature. The boundary between the dense phase and the gaseous phase is the surface of the crystal. Whenever filled cells are allowed only on top of other filled cells [the solid-on-solid (SOS) constraint] the gaseous phase is completely empty and the dense phase, the crystal, does not contain empty cells. When the cells are assumed to be cubes with only nearest-neighbor interactions, the model becomes equivalent to a Kossel crystal. At equilibrium, the (001) surface of such a Kossel crystal is rather flat at low temperatures. An edge on such a surface is then similar to the interface of a two-dimensional lattice-gas system. Here the SOS conditions can also be introduced. In such a constrained two-dimensional lattice-gas system, (1) one part of the system is completely filled, the other part is completely empty; (2) edge overhangs are excluded.

BCF [1] and Temperley [16] described the interface of a constrained two-dimensional interface model, which is in fact a SOS step model. Leamy et al. [17] showed that the same formulas can be obtained for equilibrium edges by evaluating the grand partition function. They fixed the position of one end of the interface. This corresponds physically to a step emanating from a screw dislocation, which terminates on the surface.

This model has a "critical" temperature, i.e., the temperature at which the step free energy becomes zero. At higher temperatures, however, the model becomes inconsistent, because the system can reduce its free energy by generating new steps. It is amazing that this temperature appears to be close to the critical temperature for a two-dimensional lattice gas [1]. It has been interpreted as an estimate for the transition temperature for surface roughening in a three-dimensional system. (We return to a discussion of the so-called roughening temperature in Sec. II.A.3.)

Stepped surfaces can also be simulated with Monte Carlo techniques. Recently, Leamy and Gilmer [18] studied the influence of steps on the excess surface energy. They found that the transition temperature for surface roughening is somewhat higher than predicted by two-dimensional models.

Using a Monte Carlo technique, Van Leeuwen and Mischgofsky [19] were able to bring into consideration the role of overhangs on the properties of a step. In their approach a ledge was defined by a unique path that could be followed as a line separating a higher and a lower part of the surface of the three-dimensional system. By

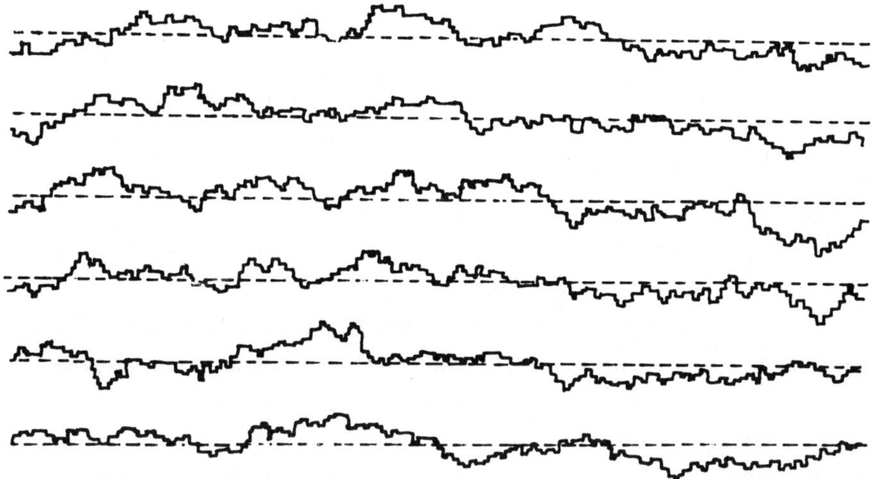

FIG. 2. Time-dependent development of microstructure of a step during a computer run of 40×10^4 exchanges at intervals of 8×10^4 exchanges; $L = 320$, $\alpha = 4.0$. (From Ref. 19.)

studying jump density distribution, jump correlations, and ledge overhang densities, they compared the properties of the BCF SOS step model, which can be calculated exactly, and the computer-generated step model, which automatically contains overhangs. In Fig. 2 the time development of a step profile in equilibrium with its vapor phase is presented. There a SOS step model is used.

The authors of Refs. 19 and 20 suggest that the dimensions of the undulations are of the order of the resolving power of modern optical techniques, such as the decoration technique used with electron microscopy. A typical configuration with a step length of 1000 units showed for $\alpha = 4$ a maximal width of about 40 units. The largest distortion was 500 units long and 30 units wide. Assuming a substance with cubic growth units of 2 Å spacing, these undulations should be visible.

2. *Two-Dimensional Critical Nucleus*

So far we have mentioned only recent work on equilibrium step structures. The statistical SOS model was generalized by BCF by imposing a supersaturation ($\Delta\mu$) on the step [1]. Then a step becomes curved and a critical two-dimensional nucleus is formed. BCF calculated the properties of such a critical nucleus. Recently, an alternative derivation of some of the BCF formulas was given using the formalism of Ref. 17, allowing the model to be extended to an anisotropic system [20,22]. Implications for the shape, size, and edge free energies were calculated and presented in an easily accessible form.

The dependence of the shape of the critical nucleus on temperature is presented in Fig. 3 for the (001) face of a Kossel crystal. At a given temperature, the dimensions of the critical nucleus change with supersaturation, but its shape does not. These calculations only hold for large two-dimensional nuclei consisting of, say, more than 100 atomic units.

3. *Roughening Temperature*

One of the most important results of recent computer simulation studies is the discovery and development of the concept of a roughening temperature [23-35]. From the point of view of crystal growth kinetics, the roughening temperature can be considered the temperature above which continuous growth occurs (the rate of growth being proportional to the driving force of crystallization $\Delta\mu$) and below which a layer growth mechanism occurs [the rate of growth being proportional to $(\Delta\mu)^2$ for a spiral growth mechanism or to $\exp(-C/\Delta\mu)$ for a nucleation mechanism].

The concept of a roughening temperature was first introduced by Burton et al. [1], where they considered the solid-vapor interface as a two-dimensional phase in which--according to Onsager [36]--a phase transition occurs. This phase transition was considered to be the roughening temperature. Applying the mean field approximation to the same two-dimensional phase, Jackson [37] introduced a recipe to calculate the α factor and used it as a criterion for the roughening temperature. This recipe was applied in a successful way to growth from the melt. In what follows, we will discuss α in more detail.

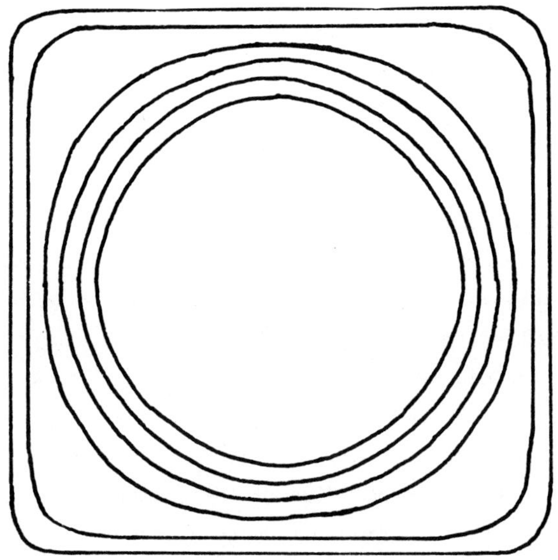

FIG. 3. The shape (not to scale) of the critical nucleus for α = 80, 40, 8, 5.28, 4, and 3.52 going from the outer nucleus to the inner nucleus. Increasing temperature results in rounding off of the corners. (From Ref. 17.)

In recent years, it was found that also for a multilayer SOS (solid-on-solid) interface model a kind of roughening temperature occurs. This whole problem is now being clarified using powerful theoretical methods like the renormalization theory developed by theoreticians in the field of critical phenomena and Monte Carlo simulations [29-35]. Very recently it has been shown that in a SOS interface, a roughening transition temperature occurs that corresponds to a phase transition of infinite order (i.e., all derivatives of the surface free energy to the temperature are zero [33-35]). It has been proved that above the roughening temperature the edge free energy becomes zero [34,35]. This predicts that above the roughening transition linear growth kinetics occur while below it growth is by a two-dimensional birth and spread mechanism.

We note that the roughening temperature is expressed either with the aid of α or by some critical "dimensionless temperature." Here α

is a factor characterizing the bond strength of the surface atoms divided by the temperature, as defined in detail later. In a recent paper on computer simulations [38], we introduced this dimensionless temperature convention and used as a unit of temperature the exact transition temperature expressed as temperature divided by the bond strength of the two-dimensional phase [1,36].

In what follows the concept of a roughening temperature will not be used explicitly. We only mention here that when a spiral growth mechanism occurs, this implies that the dimensionless "temperature" of the crystal surface is below the dimensionless roughening "temperature." This may be due to a combination of factors such as the interaction of growth units within the crystal, the interaction of the growth units in the surface with the solution or the gaseous phase, the chemical and thermodynamic properties of the solution or gaseous phase, and the temperature.

We note that so far the existence of a roughening temperature has been proven only for the (001) SOS face of a Kossel crystal together with a somewhat modified SOS model of the (110) face of a cubic bcc crystal [39]. It is reasonable, however, to assume that the concept of a roughening temperature has general validity. The precise character of the transition (transition of finite or infinite order) is not very relevant for this chapter.

4. *Physical Significance of the α Factor*

Since in our discussion the α factor plays an important role, we now discuss its significance in more detail.

So far the statistical models for surfaces and steps were considered as a part of the solid-vapor or solid-vacuum system, since the cells are either in a vapor or a solid state. For the (001) face of a Kossel crystal the α factor is defined in the following way:

$$\sigma = \frac{-4\phi_{ss}}{2KT} = -\frac{2\phi_{ss}}{KT} \tag{1}$$

where $-\phi_{ss}$ is the potential energy of a solid-solid nearest-neighbor pair and is the average bond energy per pair of interacting units. The factor 4 enters because there are 4 neighbors within a slice of a Kossel crystal. The factor α may be defined for any type of surface. This α factor can be rewritten so that we obtain Jackson's α factor [37]:

$$\alpha = \frac{\xi L}{KT} \tag{2}$$

where

$$\xi = \frac{E^{sl}}{E^{cr}} \tag{3}$$

where E^{sl} is the energy of a growth unit within a slice as defined in the Hartman-Perdok theory [40-42] and E^{cr} is the crystallization energy, i.e., the energy which is released if a growth unit enters a kink site from the gaseous phase. In Eq. (2) L is the latent heat of evaporation, T is (absolute) temperature, and K is Boltzmann's constant.

The statistical models can be generalized to a crystal-solution system by considering the vapor phase in the Ising-like models as the fluid phase, so that each cell can be either in a fluid or a solid state. The concept of a bond is then generalized and the α factor for the (001) face of a Kossel crystal can now be written as:

$$\alpha = \frac{[4\ \phi_{sf} - (\phi_{ss} + \phi_{ff})/2]}{KT} \tag{1a}$$

where $-\phi_{ss}$ is again the potential energy of a solid-solid nearest-neighbor pair, $-\phi_{sf}$ the average potential of a solid-fluid pair, and $-\phi_{ff}$ the interaction energy between the average contents of two neighboring blocks of fluid. For the case of growth from the vapor, ϕ_{sf} and ϕ_{ff} can be supposed to be equal to zero and then Eq. (1a) becomes equal to Eq. (1).

It was shown in Ref. 40 that upon introducing two assumptions we can replace (1a) by (2) also for the crystal-solution system. Now L is the "heat of dissolution" for the transformation of one solid cell into a fluid cell and T is again the (absolute) crystallization temperature.

In practice it is difficult to define an α factor for a real solid-solution interface system, since the assumptions leading to Jackson's α factor [Eq. (2)] may not hold. Moreover, most crystals consist of more than one kind of unit (ions, atoms, and complexes) and if growth takes place, they grow from a complicated solution. Much more theoretical work is needed concerning the thermodynamics of solutions in relation to the simple Ising-like models in order to clarify these problems.

In a recent paper [43] it was shown that simulations based on Ising-like models do provide an interpretation of growth from solution, and that α can be correlated with the solubility in the following way:

$$\alpha = \xi(1 - x_s)^2 (-\ln x_s + C) \tag{1b}$$

where x_s can be considered the solubility expressed in mole fraction of a saturated solution. Here C is a constant, varying from 1 to about 2.5, which depends on the difference of free energy of solid and solute particles. It can be seen from Eq. (1b) that the following qualitative role holds: the higher the solubility, the lower α.

B. Spiral Growth Models

1. *Simple Model*

It follows from the BCF nonlinear differential equation for the spiral that a spiral is almost Archimedean, i.e., after the first turn the distances between the arms of the spiral become equidistant [1, 44-47]. It follows from the analyses given in Refs. 44-47 that a unique solution exists for the differential equation when the curvature has the same sign for the whole spiral. The distance between the arms of the spiral λ_0 is given by

$$\lambda_0 = 19r^* \tag{4}$$

where r^* is the radius of a critical two-dimensional nucleus

$$r^* = \frac{\gamma}{\Delta\mu}\Omega \tag{5}$$

Here γ is the edge free energy expressed as J m^{-2}, Ω is the molecular volume, and $\Delta\mu$ is the difference in chemical potential between solute in the fluid and in the solid phase.

In the calculations mentioned above it is assumed that the critical nucleus has a circular shape. Moreover, surface diffusion and elastic stress are not taken into consideration.

2. *Back Stress Effect*

If surface diffusion is taken into consideration for a system of concentric circular steps, the so-called back stress effect occurs [48]. This means that due to the overlapping diffusion fields the center of the spiral is exposed to a lower supersaturation than the bulk supersaturation. This increases both the radius of the critical nucleus at the center and the distance between the arms of the spiral, as compared to the case where the supersaturation in the center is the same as in the bulk of the fluid.

From recent calculations, it can be shown that the back stress effect does not substantially change the shape of the ideal (pseudo) Archimedean spiral. According to Surek [48] and van der Eerden [49] the dimensions of a spiral may become much larger than the dimensions of a spiral when surface diffusion is ignored. So one may observe a beautiful Archimedean spiral when the relation $19r^*_{center}$ holds reasonably well. Yet r_{center} need not be equal to r^* (the radius of the critical nucleus), and r_{center} and consequently λ_0 may be one to two orders of magnitude larger than for a spiral free of surface diffusion or back stress.

3. *Polygonization of Spirals away from the Center*

In what follows we will discuss the phenomenon of polygonization. Polygonization means that parts of the spiral or steps are limited by more or less straight crystallographic directions.

Considering the dependence of kink density on α for the (001) face of a Kossel crystal, it follows from the computer calculations of the continuum model [46] that the spiral becomes polygonized after

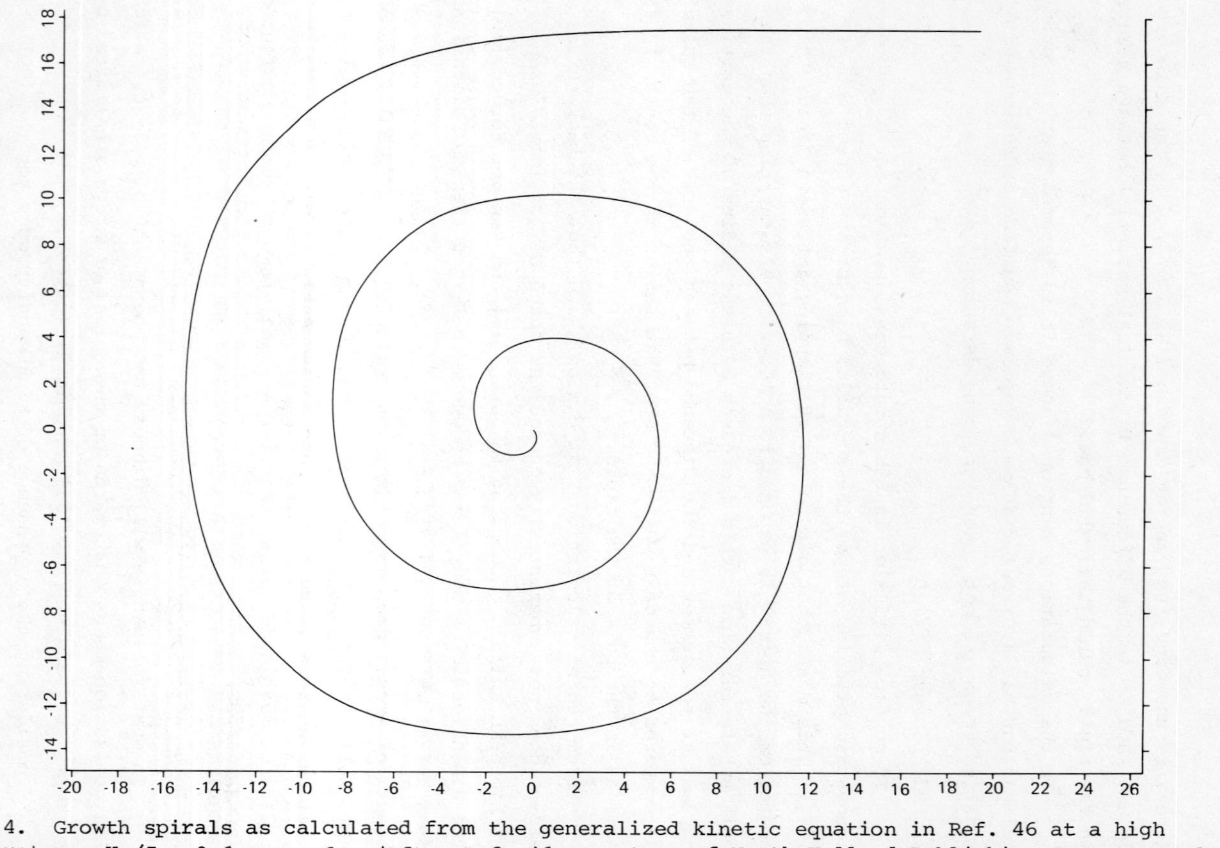

FIG. 4. Growth spirals as calculated from the generalized kinetic equation in Ref. 46 at a high temperature; $K_B/J = 0.6 \simeq \alpha = 6$. (After Ref. 46, courtesy of North-Holland Publishing Co., Amsterdam.)

the first turn for $7 < \alpha < 8$. An almost completely rounded spiral for $\alpha = 6$ (high temperature) is given in Fig. 4. A more or less polygonized spiral for $\alpha = 8$ is presented in Fig. 5, and an almost completely polygonized spiral is given in Fig. 6 for $\alpha = 10$ (low temperature). It can be seen that for a value of $\alpha = 8$ polygonization is not complete; for this case the step remains partly curved.

In the calculation mentioned above, again surface diffusion was not taken into account. Surface diffusion may give a considerable increase in the α range for which spirals become polygonized. This is because more growth units can find a kink site so that steps corresponding to a closest packed or PBC direction, having a low concentration of kink sites, show up less easily. (PBC is a periodic bond chain as defined in the Hartman-Perdok theory [41].)

We note, however, that the more there are different growth units corresponding to different chemical species in the crystal, the higher the chance that polygonization will occur. This is because then the kink sites for each particular growth unit are further apart.

If the mean displacement λ_s is smaller than x_0 (the mean distance between kinks for a certain direction), growth fronts limited by these closest packed or PBC directions show up. For the [01] direction on the (001) face of a Kossel crystal, x_0 is given by $x_0 \simeq a/2 \cdot \exp \phi_{ss}/2KT = 1/2 \cdot a \exp \alpha/4$. This means that

$$x_0 \simeq 1.1a \quad \text{for } \alpha = 4$$

$$x_0 \simeq 2.2a \quad \text{for } \alpha = 6$$

$$x_0 \simeq 3.5a \quad \text{for } \alpha = 8$$

$$x_0 \simeq 10a \quad \text{for } \alpha = 12$$

Quantitative studies concerning the influence of surface diffusion on shapes of spirals have not been carried out.

It can be concluded from the discussion given above that the higher the mean displacement λ_s the higher the value of α for which polygonization off the center occurs.

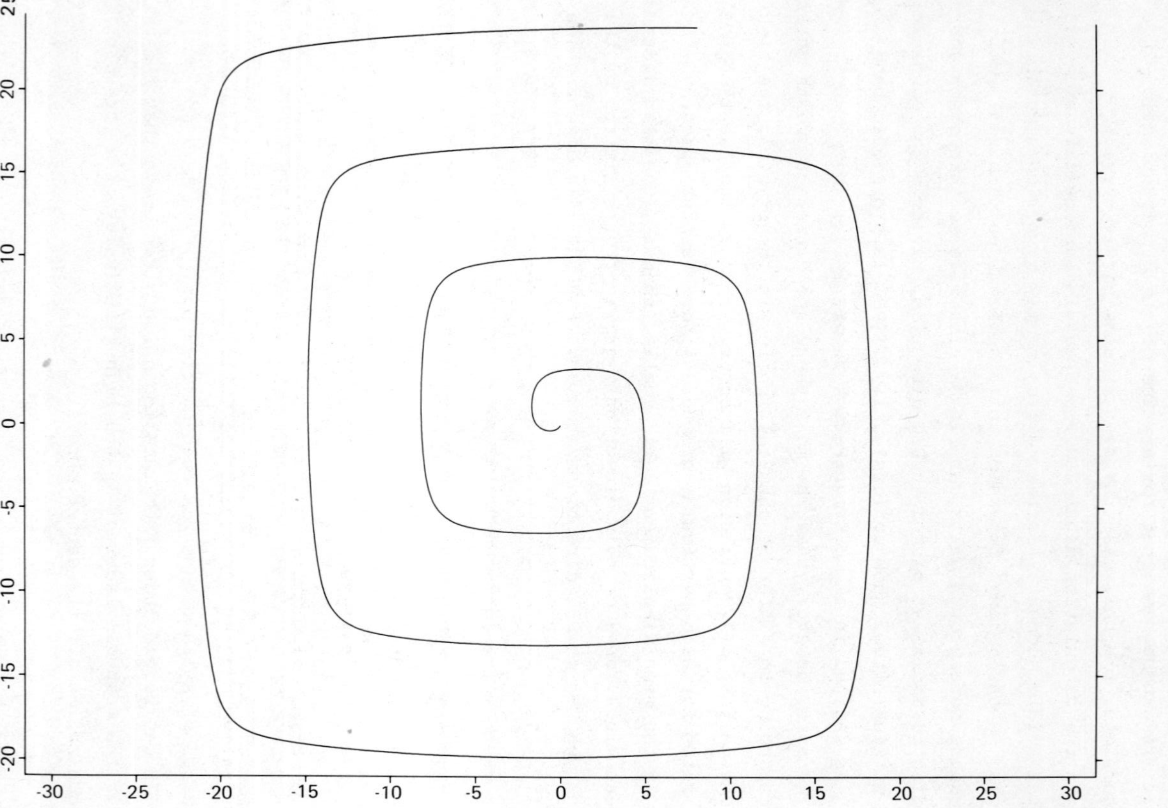

FIG. 5. Growth spiral at an intermediate temperature. The spiral becomes polygonized due to increasing anisotropy of both kinetic coefficient and edge free energy; $K_B/J = 0.56 \simeq \alpha = 8$. (After Ref. 46, courtesy of North-Holland Publishing Co., Amsterdam.)

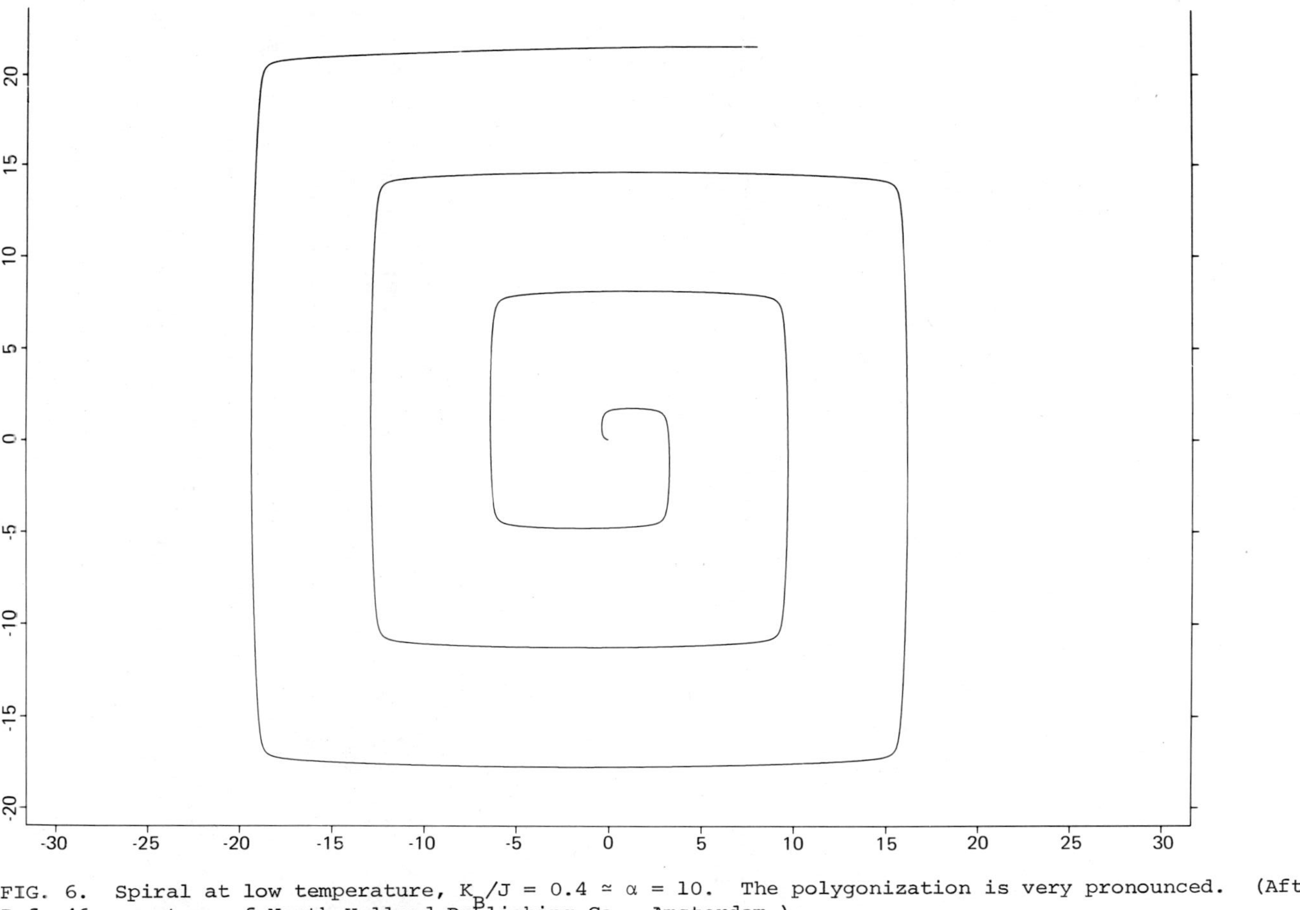

FIG. 6. Spiral at low temperature, $K_B/J = 0.4 \simeq \alpha = 10$. The polygonization is very pronounced. (After Ref. 46, courtesy of North-Holland Publishing Co., Amsterdam.)

It was also found by Müller-Krumbhaar et al. [46] that polygonization gives a reduction of the distance between the arms of the spiral compared with an ideal rounded, pseudo-Archimedean spiral. Instead of a value of $\lambda_0 = 19r^*$ [Eq. (4)] the value of λ_0 for a more or less complete polygonized spiral such as given in Fig. 6 is approximately given by

$$\lambda_0 \simeq 4\sqrt{2}\, r^* \cong 5.4r^* \tag{4a}$$

These results are in agreement with the analyses of Kaishev [50] and Boistelle and Doussoullin [51].

4. Polygonization in the Center

It follows from previous calculations [20,22] that for $25 < \alpha < 40$ the shape of the critical nucleus gradually changes from a circular shape to an (almost) square shape (see Fig. 2) as one goes from a low α value (high temperature) to a high α value (low temperature). Since the curvature of the center of the spiral is determined by the critical nucleus if surface diffusion is not taken into consideration, this implies that polygonization of the center occurs at much higher values of α (lower temperatures) than polygonization away from the center.

5. Monte Carlo Simulation of a Spiral

Very recently Monte Carlo computer simulation experiments were carried out to simulate spiral growth on the basis of an atomistic model [26,52]. Especially for low α values, these studies are of great importance for checking the detailed atomistic structure of the spiral center and spiral arms, i.e., the roughness of the spiral (see also Sec. III for the roughness of steps). These studies are also very important for checking the domain of validity of calculations carried out for condinuous models.

Swendsen et al. [52] used a computer program which enabled them to study spiral growth as the dominant process, with two-dimensional nucleation playing a negligible role. Thus conditions were fulfilled for a sufficiently high α factor--or sufficiently below the roughening temperature corresponding to $\alpha \simeq 3.2$--and small values of $\Delta\mu/KT$, which is roughly equal to the relative supersaturation. Therefore, these authors neglected nucleation by excluding the condition of isolated particles or surface vacancies. This condition substantially reduces the set of possible creation and annihilation sites at any given time. In their program, the authors kept track of these sites and their surroundings (number of bonds), in addition to the crystal surface. Since the time unit of the simulation is determined by the fastest rate in the problem, the speed of the program was increased by a factor of $e^{1/4\alpha}$ due to the elimination of the creation and annihilation of isolated single particles and vacancies. This decreased the necessary computer time by two to three orders of magnitude. In Fig. 7A and B, two typical spirals are presented for $\alpha = 10$ and $\beta = \Delta\mu/KT = 0.6$, and for $\alpha = 4$ and $\beta = 0.4$. Figure 8A and B also show another Monte Carlo simulation of spirals, reproduced from Ref. 29.

Upon comparing the spiral of Fig. 7A obtained from a Monte Carlo calculation for $\alpha = 10$ with the spiral obtained from a calculation of a continuous model for $\alpha = 8$ and $\alpha = 10$ given in Figs. 5 and 6, it can be seen that the correspondence is good, apart from the roughness in Fig. 7. It is indeed found by analyzing the results of the calculations of the continuous model [46] and of the discrete atomistic model [49] that the results of the two approaches are in excellent agreement. Thus the Monte Carlo simulations confirm to a great extent the properties of the spiral discussed in Section II.B.1. Also Eq. (4), concerning the distance between the arms of a spiral, is confirmed to a great extent.

We mention that Gilmer [26] has also recently carried out Monte Carlo simulations of the spiral growth process. A computer program was used without the restrictions of Ref. 45. Again it was found that the results of the Cabrera-Levine theory hold amazingly well, notwithstanding the fact that the presuppositions of their theory may not

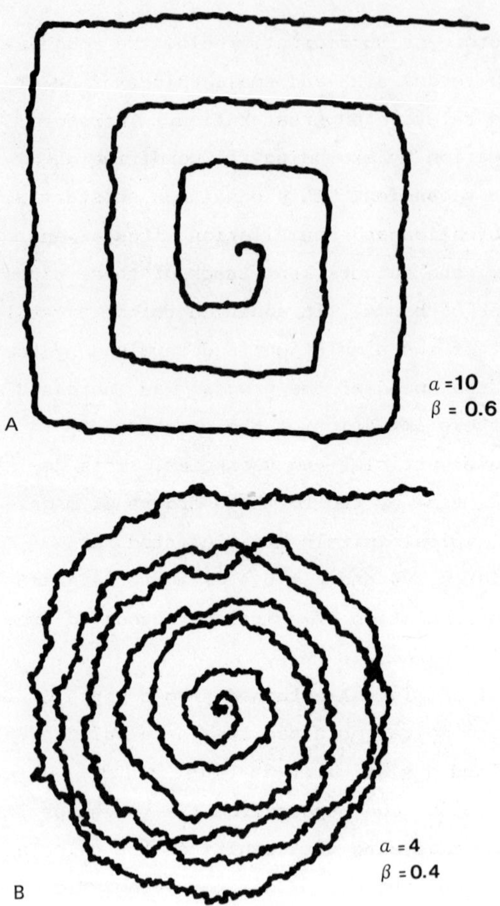

FIG. 7. Monte Carlo simulation of a spiral: (A) $\alpha = 10$, $\beta = \Delta\mu/KT = 0.6$; (B) $\alpha = 4$, $\beta = 0.4$. (From Ref. 52, courtesy of North-Holland Publishing Co., Amsterdam.)

hold. For an α value of 8 and a value of $\Delta\mu/KT = 1.5$ a double spiral originating from two screw dislocations with a Burgers' vector of 2a turned out to be quite rough. The same holds for an α value of 12 and $\Delta\mu/KT = 3$. For α values of 20 and 16 and $\Delta\mu/KT = 3$ polygonization could be observed. It is interesting to note that a spiral for a high α value may become very similar to a spiral at a low α value if the $\Delta\mu/KT$ becomes quite high. This is also in agreement with the Cabrera-Levine theory, as can be seen from Eq. (5).

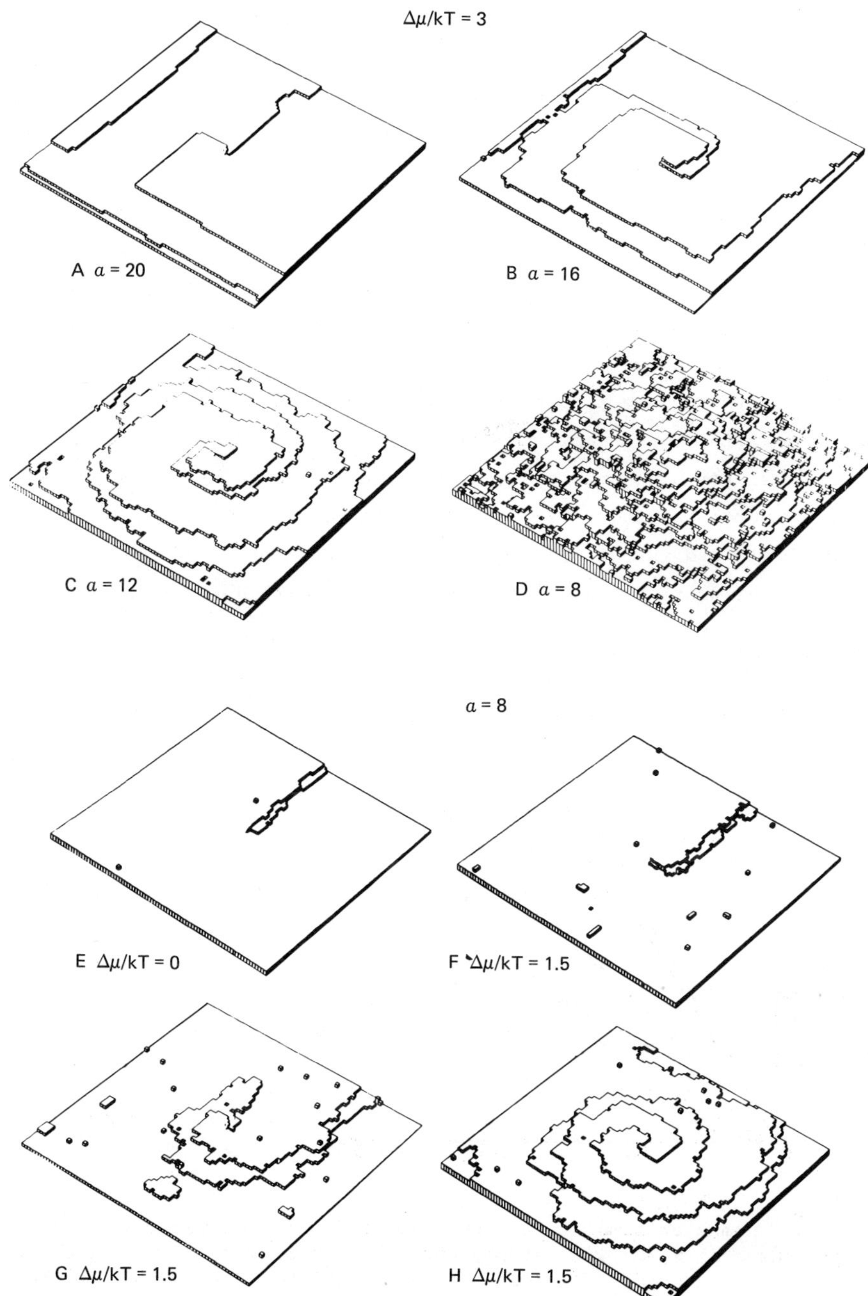

FIG. 8. Monte Carlo simulation of double spirals at a constant $\Delta\mu$ /KT and different α values (top four); and at a constant α but different $\Delta\mu$/KT (lower four). (After Ref. 29, courtesy of North-Holland Publishing Co., Amsterdam.)

The distance between the arms of the spiral λ_0 is determined only by $\gamma/\Delta\mu$. If α decreases, γ also decreases and this may be compensated for by a decrease in $\Delta\mu$.

6. *Conditions for the Occurrence of a Rough Spiral: Spiral Growth and Two-Dimensional Nucleation*

In Fig. 9 the normalized measured growth rate R versus β (roughly relative supersaturation) is presented for growth with and without steps for two values of α ($\alpha = 3$, $\alpha = 4$). For $\alpha = 3$ the surface is above the roughening temperature and is so rough that introduction of steps hardly gives an increase in growth rate (see Fig. 9, curves a and b). For $\alpha = 4$ the surface is below the roughening temperature and is much less rough. It was shown that a two-dimensional nucleation, birth and spread model gives a good description of the Monte Carlo data, where the edge free energy per growth unit γ is given by

$$\frac{\gamma}{KT} = 0.49$$

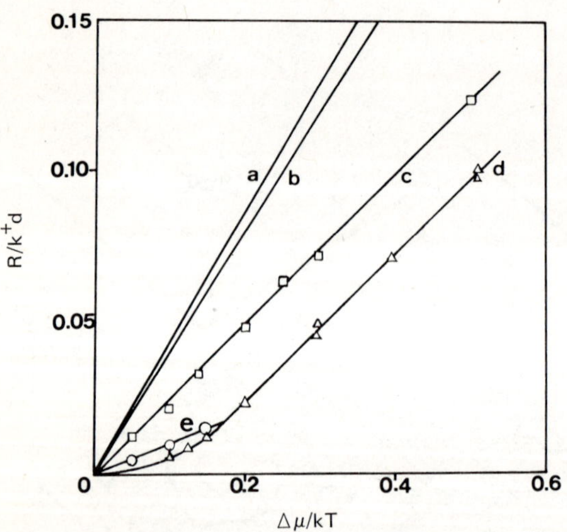

FIG. 9. Computer simulation curves for a (001) face of a Kossel crystal: curve a, $\alpha = 3$ with steps; curve b, $\alpha = 3$, without steps; curve c, $\alpha = 4$, with steps (□); curve d, $\alpha = 4$, without steps (Δ); curve e, $\alpha = 4$, with a five times lower step density than curve c (O). For values of $\beta = 0.2$ a compound mechanism occurs (see curve e). R/k^+d is the normalized growth rate, while $\Delta\mu/KT$ is driving force for crystallization, which is roughly proportional to the supersaturation. (After Ref. 23, 24, courtesy of North-Holland Publishing Co., Amsterdam.)

(see curve d in Fig. 9). The value of γ will be somewhat lower if better formulas are used (Ref. 20) but this does not change our argument. It will be shown in Sec. III that a growth spiral with a rough step is occasionally observed. This suggests a low α value. A low α value, however, implies that nucleation already occurs at low supersaturation. We therefore will discuss conditions under which a "rough spiral" may develop.

If a high concentration of steps is introduced corresponding to an average distance between the steps of 5 interatomic distances, the step mechanism dominates the nucleation mechanism (curve c in Fig. 9). If, however, a system of steps is introduced with a 5 times lower step density corresponding to an average step distance of 25a, it can be seen from the curves e and d in Fig. 9 that for values of $\beta \simeq 0.2$ a compound mechanism operates. A compound mechanism is a mechanism where both spiral growth and two-dimensional nucleation occur simultaneously. We note that the distance 25a corresponds to the spiral of Fig. 8. For $\beta \lesssim 0.2$ the screw dislocation mechanism dominates. For values of $\beta \gtrsim 0.2$ the two-dimensional birth and spread mechanism is dominant (see curve c).

Upon substituting the values $\gamma/KT = 0.49$ and $\Delta\mu/KT = 0.4$ in Eq. (5) (Sect.II.B.1) it is found that

$$r^* \simeq 1.2a$$

This means that the radius of the critical nucleus is about 1 atomic unit and that a critical nucleus for $\alpha = 4$ and the given $\Delta\mu/KT$ contains about 3 or 4 atoms. Although Eq. (5) is not applicable for very small nuclei (this formula only holds for a large nucleus with a well-defined circular shape) it nevertheless gives a good approximation of the atomic dimensions of the "critical like nuclei."

In this connection it is interesting to note that according to the pair approximation model of Gilmer et al. [53] for $\alpha = 4$ and $\beta \gtrsim 0.5$, the so-called pair approximation model applies. This suggests that for these conditions the critical nucleus is somewhat larger than 2 units.

It is also interesting to note that Eq. (4) for the distance between the arms of the spiral for $\lambda_0 = 19r^*$ (see Section II.B.1) holds roughly since $\lambda_0 \simeq 27a$.

In the computer program of Gilmer and Bennema [23,24] leading to the (R,β) curves given in Fig. 9, the chances for creation (condensation) on the 5 possible sites of a Kossel crystal were chosen to be equal. Thus contrary to the computer program of Swendsen et al. [52] leading to the spiral of Figs. 7 and 8, the creation (and evaporation) of atoms on a flat surface and the annihilation (and creation) of surface vacancies were not suppressed. This leads to the observation that for the simulation of Gilmer and Bennema a two-dimensional nucleation mechanism is a dominant mechanism for the very small nuclei for $\beta > 0.2$ and a step density corresponding to $\lambda_0 \simeq 25a$. In the simulation of Swendsen et al. the artificial suppression of two-dimensional nucleation causes a rough spiral.

In Fig. 10 normalized rates versus supersaturation curves are presented for spiral growth, two-dimensional nucleation without a spiral and spiral growth in which two-dimensional nucleation is suppressed. This is the result of recent Monte Carlo simulations

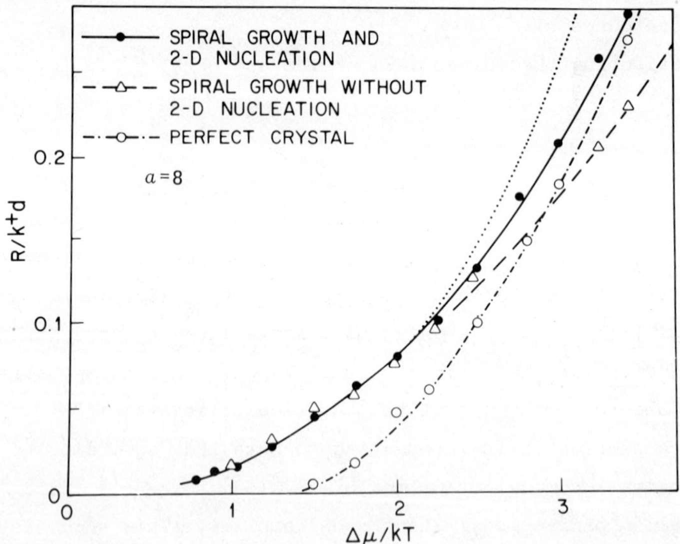

FIG. 10. Normalized rate versus supersaturation curve obtained by computer simulation. (After Ref. 26, courtesy of North-Holland Publishing Co., Amsterdam.)

carried out by Gilmer [26]. It can be seen that for this high α value of $\alpha = 8$, two-dimensional nucleation becomes dominant for $\Delta\mu/KT > 3$. The parabolic $(R, \Delta\mu/KT)$ law as predicted by the BCF theory is indeed more or less found from the computer simulations, as can be seen from Fig. 10. In fact, the probability that a spiral as given in Figs. 7 and 8 may really occur seems to be very low. However, three things may occur and give rise to a rough spiral:

1. A group of about five dislocations with equal sign, with a small separation of about $\lambda_0/2 = 27a \simeq 13a$, emerge on the surface within a circular area with radius of 13a. Here we use the spiral of Fig. 7 as a reference. Then a group of cooperating spirals occurs which reduces λ_0 to $\lambda_0 \simeq 5a$.
2. The exchange kinetics between growth units in the fluid phase and units on the free surface in the adsorption layer (or in the surface) may be very slow as compared to the exchange kinetics between growth units in the fluid phase and in the steps. This may occur due to a very strong adsorption on the free surface by an impurity or the solvent.

   Then two-dimensional nucleation is suppressed in the same way as in the computer program of Ref. 46. (In Ref. 46 this program was not used to save computer time; only in this way could spirals with a sufficient number of turns be produced.)
3. Another reason may be the occurrence of surface diffusion. If the mean displacement λ_s caused by surface diffusion is larger than λ_0 [Eq. (4)], the formation of two-dimensional nuclei between the arms of the spiral is again suppressed. On the other hand, if we assume that the new spiral theory of van der Eerden also applies to these rough spirals, a large λ_s makes the distance between the arms of the spiral larger [49].

For an average step density corresponding to curve c in Fig. 9 ($\lambda_0 \simeq 5a$), fragments of the roughened steps collide during the growth process and form fragments of double steps. Then fragments of double steps move much slower than the other pairs of the steps, so that if growth progresses, the step patterns become continuously rougher. This is demonstrated in Fig. 11.

3	4	3	3	3	3	4	4	4	4	4	5	5	5	4	4	4	5	5	5
5	4	4	3	3	3	3	4	5	5	5	5	5	5	5	5	5	5	5	5
5	4	3	3	4	5	5	5	6	6	4	6	5	5	5	5	5	5	5	5
5	4	4	4	4	5	5	6	6	6	6	6	5	5	5	6	5	5	5	5
5	4	5	5	5	5	5	6	6	5	5	6	3	5	5	6	6	5	5	5
5	5	5	5	5	5	6	6	7	6	6	6	5	5	5	6	6	6	5	5
5	5	5	6	6	6	6	6	6	6	6	7	6	6	6	6	6	6	5	5
5	5	6	6	6	6	6	6	6	6	6	6	6	6	6	6	6	6	5	5
6	5	6	6	6	6	6	6	6	6	6	6	6	7	6	6	6	6	5	6
6	6	7	6	6	6	6	8	7	7	6	6	6	6	6	6	6	5	6	6
6	7	7	7	6	6	7	7	7	7	7	6	6	6	7	6	6	6	6	6
7	7	8	6	8	7	7	7	7	7	7	6	6	6	6	7	7	7	7	7
6	8	8	8	6	7	7	6	8	7	7	6	6	6	6	7	7	7	7	7
7	8	8	8	7	8	8	8	8	7	7	7	7	7	7	6	7	7	7	8
8	8	8	8	9	9	9	8	8	7	7	7	7	8	7	7	7	7	8	8
8	8	8	8	9	9	9	9	8	8	8	8	9	7	7	7	8	8	8	8
8	8	8	9	9	9	9	9	8	9	8	9	9	7	7	7	8	8	8	8
9	9	8	9	9	9	9	9	9	9	9	8	9	8	6	8	8	8	8	8
9	9	9	9	9	9	9	9	9	9	9	9	9	8	7	8	8	8	8	8
9	8	8	9	8	8	9	9	9	9	9	10	10	10	8	8	8	8	9	8

FIG. 11. Surface configuration with five monoatomic steps as printed by the computer. The numbers represent the height of the crystal surface above some arbitrary reference level, and the dark lines represent the positions of the steps. This was printed during growth with $\alpha = 4$ and $\beta = \Delta\mu/KT = 0.3$. (After Ref. 23, courtesy of North-Holland Publishing Co., Amsterdam.)

Such a kinetic roughening can also be seen in the spiral of Fig. 7, where notwithstanding the much lower step density as compared to Fig. 11 already after a few turns three "contacts" are formed. It is expected that if a rough spiral occurs, the spiral becomes rougher the further it is from the center. When the step density is high due to high supersaturations and/or a high number of cooperating spirals, this "roughening" process goes faster. This kinetic roughening process must be distinguished from the thermal equilibrium roughening process mentioned in Sec. II.A.3. The kinetic roughening process leads to perturbations which are longer and have a higher amplitude than the equilibrium perturbations.

So far we discussed the occurrence of rough spirals formed under conditions of high supersaturations. Such spirals are the only spirals which can be simulated. If we take a value of $\beta = \Delta\mu/KT$ lower by 100 times than the value of $\beta = 0.4$ corresponding to the spiral of Fig. 7, we get [according to Eqs. (4) and (5)] a spiral with the same shape as the one shown in Fig. 7. However, the distance between the arms of the spiral becomes 100 times larger, so that $\lambda_0 \simeq 27 \times 100a = 10^4$ Å, where we assumed $a \simeq 3$ Å. This means that the distance between the arms of the spiral is then a few μm. For such low supersaturations two-dimensional nucleation between the arms of the spiral does not occur. This is independent of the surface kinetics. The chances for the occurrence of kinetic roughening are also much less than for spirals with a high density due to the higher step distances. However, kinetic roughening may occur if a group of cooperating spirals occurs in the growth center or if a spiral is emitted from a screw dislocation with multiple Burgers' vector.

An example of kinetic roughening due to a double step emitted from a screw dislocation with a Burgers' vector having a component of 2a is given in the paper of Gilmer [26].

7. *Multimolecular Spirals*

So far only spirals with a height of 1 unit were discussed. According to the BCF theory, the distance between the arms of a spiral is given by Eqs. (4) and (5). In this theory the height of the step is not specified. In forthcoming papers [47,56], we will show that from the generalized Cabrera-Levine theory (in which the strain due to a high Burgers' vector is introduced) the shape of a spiral with a step height of say 500 to 1000 Å can be derived, which is in perfect agreement with observed spirals. It is then interesting to compare the shapes of spirals which grow under the same supersaturation on crystal surfaces, but which differ in step height, due to a difference in the Burgers' vector of the dislocation from which the spiral step originates.

In principle the edge free energy of a monomolecular step can be calculated for simple Kossel-like crystals, as discussed in Sec.

II.A.1. It is given by the separate rule $\gamma/KT = \alpha/4 - \ln(\cosh \alpha/8)$ (see Refs. 17, 20). For multimolecular steps, however, no exact calculations are possible and good approximations have not yet been derived. In the future it is necessary to investigate the edge free energy as a function of the step height, under conditions for which a macrostep is stable, under conditions for which it decomposes, etc. In the case of a monomolecular step with a [100] direction on the (001) face of a Kossel crystal, the edge free energy per growth unit in the edge is $\alpha/4$, if we neglect the entropy of the roughening term in the equation for γ given above. If the step becomes n times higher, the edge free energy per unit surface becomes somewhat larger due to a decrease of the entropy. If the roughening entropy of a step is taken into consideration, the edge free energy is reduced for a monomolecular step, which for low α values (high temperatures) may be quite high as discussed above. The higher the steps the less the influence of this roughening, due to the fact that the average potential energy of the blocks in a macrostep is much lower since they have more bonds on the average.

Thus if we compare a monomolecular step with a multimolecular step, it can be estimated that due to the difference in edge entropy the edge free energy of the monomolecular step may be 2 or perhaps 4 times lower than the multimolecular steps. If we compare multimolecular steps with different step heights, these differences are much less. Assuming a relation $\lambda_0 = 19r^*$ [Eq. (4)], which holds independent of the step height if we neglect the influence of stress, we may expect from Eq. (5) that the distance between the arms of a monomolecular spiral may be 2 (to 4) times lower than the distance between the arms of two-molecular spiral. Distances between multimolecular spirals with steps of a different height are probably much less. If surface diffusion is taken into consideration, the differences mentioned above probably remain valid notwithstanding the fact that Eq. (4) is no longer valid.

So far we were only comparing monomolecular spirals with multimolecular spirals for which the steps are real macrosteps, which for a Kossel crystal have a 90° profile. If a macrostep decomposes slightly so that the profile deviates slightly from 90°, the differences between macroatomic steps and multiatomic steps may be very small due to the edge entropy.

So far polygonization was not taken into consideration. The higher the step and the more ideal the macrostep the more the chance that polygonization occurs. It may be expected that in the same way as for polygonized monomolecular spirals the distance between the arms of the spiral is reduced if polygonization occurs.

We note that a theory was developed by Cabrera concerning the advance velocity of steps of a spiral resulting from a screw dislocation with a very high Burgers' vector [54]. Cabrera assumed that two-dimensional nucleation occurs on the (flat) steps above a certain supersaturation. This is because above a certain (not very high) supersaturation the macrosteps can move faster than steps with a much smaller height, since the nucleation barrier for nucleation against the step is reduced considerably (due to the reentrant corner between the flat macrostep and the surface).

It is interesting to note that for sufficiently high α values, monomolecular steps are limited by periodic bond chains (PBCs) which occur in the slice of the F face (hkl) under consideration. These have a thickness $d_{nh\ nk\ nl}$ determined by the elementary cell and space group. Spirals limited by high multimolecular steps may be limited by macrosteps having an orientation of an F face of which slices consist of at least two sets of different PBCs. We use here the concepts of the Hartman-Perdok theory [41,42].

8. *Spirals Growing under the Influence of Strain*

So far we have only discussed normal steady-state growth spirals. Following Cabrera and Levine [44] it will be shown [47,55,56] that upon introducing a homogeneous stress field in the center of a spiral the positive curve may become much less positive or even weakly negative. If negative, this gives rise to a kind of depression, especially for large groups of cooperating spirals. A screw dislocation with a high Burgers' vector gives rise to an elastic stress field with a steep gradient close to the origin, causing development of a hollow core with a diameter in the μm range in the center of the growth spiral. It follows from the theory that the spiral step escapes from this hollow core after passing a turning point at which the curvature changes from a negative to a positive curvature. It will be shown that the hollow core and the change of curvature are indeed found experimentally [47,56].

9. *Dissolution Spirals*

The whole theoretical treatment also applies to dissolution spirals. Upon changing from a certain supersaturation to the same absolute, but negative, supersaturation (or undersaturation) the steady-state spiral shape must be the same, except that now a negative spiral develops into an etch pit. This symmetry between growth and dissolution spirals does not hold if strain is introduced. Cabrera and Levine showed [44] that in this case a negative spiral with a hollow core may develop; contrary to the case of growth, however, no turning point occurs, i.e., the negative spiral will have a reversed trumpet shape. If the undersaturation surpasses a certain value, the hollow core opens up so that a macroscopic pit develops [44,47,56].

In general, however, the surface topology for growth and dissolution are not symmetrical, because in dissolution many more sites of attack are possible as compared to growth sites, e.g., dissolution at edges and edge dislocations.

10. *Non-Steady-State Growth versus Steady-State Growth; Gradient in Supersaturation*

So far we only discussed steady-state growth and dissolution spirals. With a group of cooperating spirals very complicated non-steady-state spirals may occur, if the distance between the points of emergence of the screw dislocations is smaller than $19r^*/2$ (if no surface diffusion occurs; see Refs. 1, 40). Sometimes, however, pseudoperiodic steady-state "rhythmic bunching" phenomena may be imposed on a spiral [57], although this usually leads to non-steady-state phenomena.

A steady-state growth spiral shows a weak upward curvature in the center. With a sudden increase in supersaturation this upward curvature becomes much more pronounced, due to a tighter winding of the spiral arms. With a decrease, the summit of a spiral becomes rounded off as compared to the ideal steady-state spiral.

When a linear gradient of supersaturation is imposed over the surface with a spiral system, an eccentric steady-state spiral may develop with a wider spacing between the steps on the high supersaturation side and a lower spacing on the lower supersaturation side [58]. Complicated patterns may occur if several gradients in supersaturation are imposed on a spiral.

11. *Spiral Growth and Growth by a Two-Dimensional Birth and Spread Mechanism*

The competition between spiral growth and two-dimensional nucleation growth was considered in Ref. 26. As can be seen in Fig. 12 for a value of $\alpha \simeq 8$, above $\Delta\mu/KT \simeq 2.7$ a two-dimensional birth and spread mechanism is dominant while below this value a spiral growth mechanism predominates when a screw dislocation is introduced on the surface in the computer program.

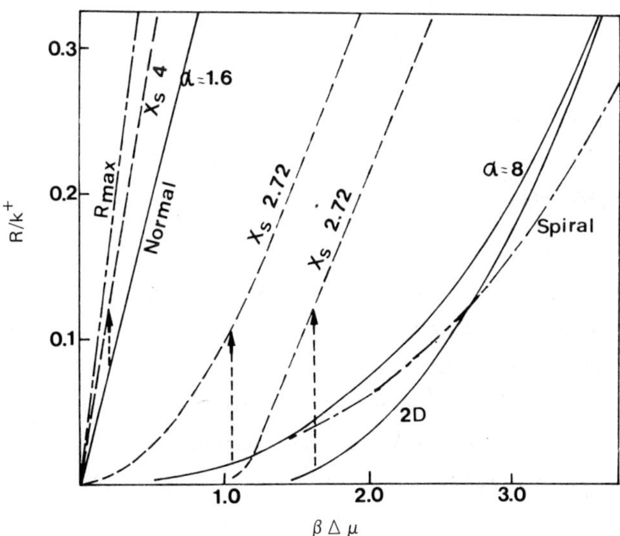

FIG. 12. The influence of surface diffusion on the three different growth mechanisms of a Kossel (100) face. Data at $\alpha \simeq 1.6$ from Ref. 24, at $\alpha \simeq 8$ from Ref. 26. (After Ref. 38.)

The transition from spiral growth to two-dimensional nucleation growth depends on α: the higher the α the higher the $\Delta\mu/KT$ range for which spiral growth occurs. This also depends on surface diffusion, with the higher the mean displacement caused by surface diffusion, the lower the $\Delta\mu/KT$ value for which this transition occurs. Using the intuitive concept of a σ^* which marks the transition from spiral growth to two-dimensional nucleation growth, we can say that the higher α, the higher is σ^*, while the larger is the mean displacement, the lower is σ^*.

12. *Role of Impurities*

The role of impurities has not yet been mentioned. In general, impurities must be included in crystal growth models. They may reduce the edge free energy and hence decrease the distance between the arms of the spiral [see Eqs. (4) and (5)]. They may block kink sites and therefore reduce the distance between kinks which can absorb growth units, which may make $\lambda_s < \lambda_0$ and hence increase the probability of polygonization. If the surface is dotted with fixed impurity molecules, steps may have difficulty in pushing through the impurities. If the distance between impurity molecules is larger than r^*, parts of the step may go through while only portions are blocked (see Ref. 59). This gives a very rough step profile and a spiral with a rough step and a positive curvature.

With dissolution, impurities may block the backward movement of the steps and thereby cause negatively curved steps to develop.

13. *Comparison of Growth from the Vapor and Solution*

In Sec. III we will show spirals on crystal surfaces grown under a variety of conditions, e.g., from a pure vapor, transport by chemical vapor, aqueous solution, and from fluxes. Therefore, we now try to derive some criteria concerning the difference between the spiral shapes on crystals grown from these different phases.

Since so many unknown factors are involved in these complicated growth processes, clear-cut general conditions cannot be specified. First we assume for the time being that for growth from the vapor the α factor is usually higher than for growth from a solution, due to the fact that in solution growth the ϕ_{sf} bonds compensate to a high extent the ϕ_{ss} bonds because there may be a strong interaction between the solvent and the units on the surface [see Eq. (1a)]. This overcompensates the possibly higher temperature for vapor growth compared to solution growth. With a higher α factor, the following conclusions can be drawn:

1. Spirals on crystals grown from the vapor are more polygonized than spirals on crystals grown from solutions.
2. Spirals on crystals grown from the vapor have a larger step distance than spirals grown from solution because $\gamma_{vapor} > \gamma_{sol}$.
3. Spirals on crystals grown from the vapor may be less rough than spirals grown from solution because of the higher α values.

However, conclusions like the above have a limited value. As an example we mention that it was found from an analysis of measure rate versus supersaturation for growth from solution that the mean displacement λ_s was about 10a [60], although occasionally it was one order of magnitude higher. This is probably less than the mean displacements for growth from the vapor. Thus in growth from the vapor, polygonization is suppressed compared to growth from solution. This argument weakens statement 1. On the other hand, the larger the mean displacement the larger the distance between the arms of the spiral [48,49]. We note that upon comparing different spirals occurring on crystals grown under totally different conditions the supersaturation may not differ too much.

14. *Summary of the Properties of a Spiral as Follows from Modern Theories: Applicability of Models*

We discussed the results of simple models such as a Kossel crystal. If, for example, next-nearest-neighbor interactions were included in the Kossel crystal, the shapes of the polygonized spiral could become octagonal. However, such phenomena can be accounted for by using the concepts of the Hartman-Perdok theory. No sophisticated treatment has been carried out on more complicated crystal structures for which the most fascinating spirals are actually observed. For the time being we can only try to extrapolate the results of continuous, statistical mechanical and computer simulation models to more complicated structures in a qualitative way. We then arrive at the following conclusions:

1. The symmetry of the crystal face is more likely to exhibit itself in polygonized spirals as one increases the ratio of the crystallization temperature to the effective bond strength ($[\phi_{sf} - (1/2)(\phi_{ss} + \phi_{ff})]$; [see Eq. (1a)]), increases α, lowers the solubility, or lowers the surface diffusion.
2. Steps on surfaces characterized by rather low α values ($3 \lesssim \alpha \lesssim 7$) show roughness due to thermal fluctuations, which in principle are observable with modern electron optical techniques. Perturbations having a width of about 30 to 40 units and a length of about 500 units are expected for an α of about 4.
3. Ideal stationary spirals have an almost Archimedean shape, i.e., after the first turn they become equidistant with a distance between the arms of a spiral of $19r^*$ (r^* being the radius of the critical nucleus). With surface diffusion and a mean displacement $\lambda_s > 19r^*$, the distance between the arms of the spiral increases; the pseudo-Archimedean character, however, probably will not change (this is due to the so-called back stress effect).
4. Polygonization away from the center occurs for α larger than about 7, provided that no surface diffusion occurs. With surface diffusion, this critical value of α may go up considerably. Polygonization reduces the distance between the arms of the spirals from $19r^*$ to about $5.4r^*$. Also with surface diffusion polygonization may occur provided that $\lambda_s < x_0$ (x_0 is the mean distance between kinks in a straight step).
5. Polygonization in the center occurs in general at much higher values than polygonization of the arms of the spiral.
6. Only spirals with a rough step profile occur if α is small. In spirals with a high step density the supersaturation is high and two-dimensional nucleation must be suppressed. The further from

the center, the rougher is the spiral because of the kinetic roughening process which occurs for rough steps. This only holds if step distances are sufficiently small.

7. Spirals with multimolecular steps are more polygonized than spirals with monomolecular steps; the step distance is two times larger for multimolecular steps as compared to monomolecular steps.
8. As a tentative conclusion it can be stated that spirals of crystals grown from the vapor may show more polygonization and a higher step distance than spirals of crystals grown from solution. This statement only holds if mean displacements and supersaturation do not differ too much in these two situations.
9. Impurities may drastically alter the shapes of spirals. Step distances may be reduced and step profiles may become rougher.
10. A flat stress field in the center of a spiral introduces a weak positive or negative curvature. A steep local stress field caused by a dislocation introduces a hollow core.
11. Cooperating groups of spirals may introduce rhythmic bunching. Gradients in supersaturation cause eccentric spirals. The higher the local supersaturation the larger is the local distance between turns of the spiral.
12. Both two-dimensional nucleation and spiral growth may occur simultaneously for low α values, both for growth and dissolution, for certain small ranges of the supersaturation.

III. EXPERIMENTAL RESULTS

A. Two Types of Crystal Faces

Two essentially different types of surface microtopographs can be distinguished when crystal faces are observed using an optical microscope with incident lightning.

On most well-developed crystal faces, one can see surface microtopographs which resemble contour lines on a geographic map. These correspond to growth steps developing on the surface, and may be irregular, circular, or polygonal. When the separations between the

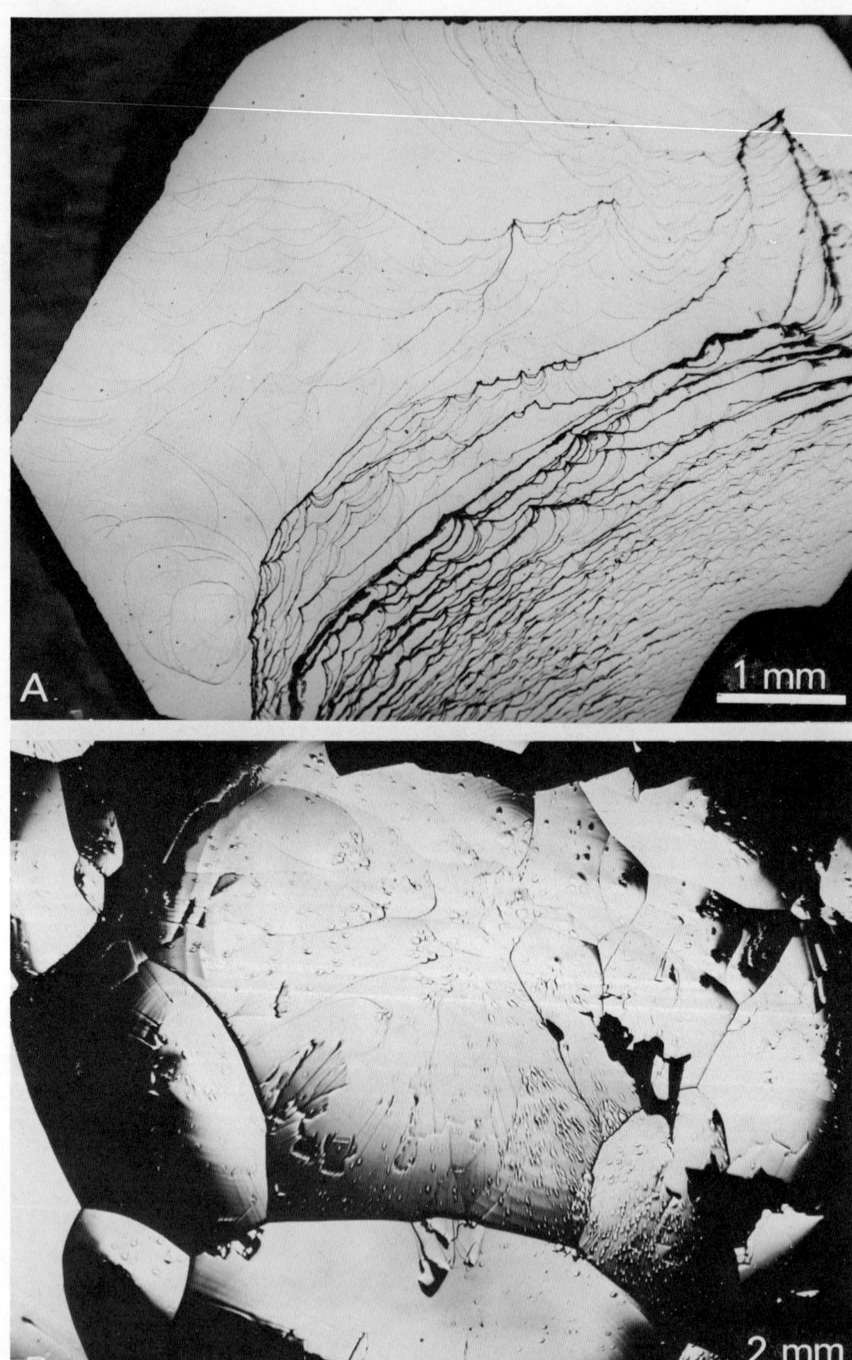
A
1 mm
B
2 mm

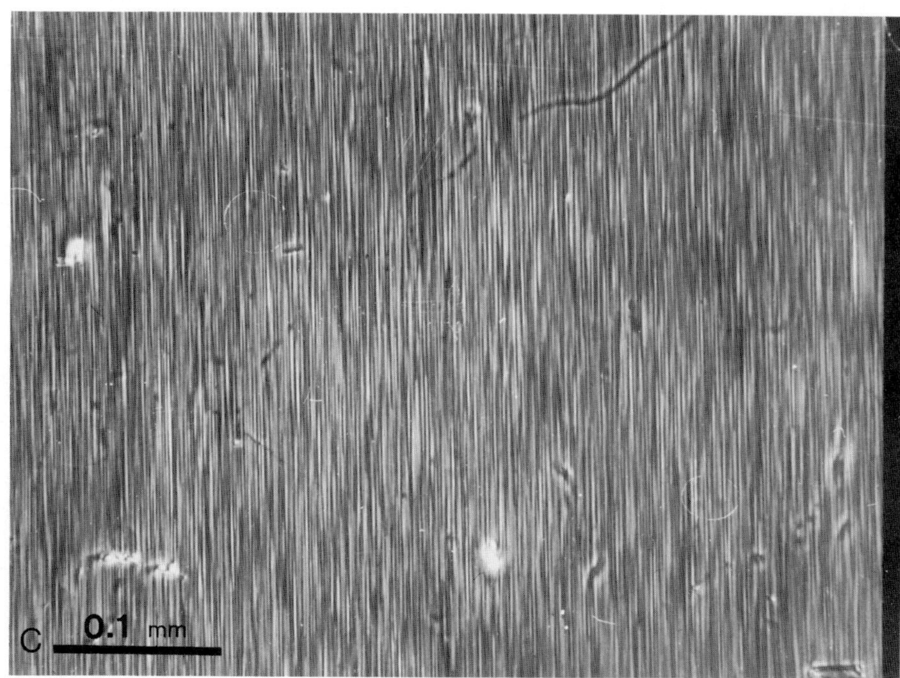

FIG. 13. Three representative surface microtopographs of crystal Ref. 208.) Three representative surface microtopographs of crystal faces: (A) Well-developed crystal face showing a step pattern; natural hematite, (0001); ordinary reflection photomicrograph. (B) Well-developed crystal face with a large number of growth hillocks; under higher magnification, these hillocks are seen to consist of narrowly spaced growth steps; synthetic quartz, minor rhombohedral face; ordinary reflection photograph. (C) An example of crystal faces showing only striations; even under the highest magnification, no growth layers can be seen on this type of face; natural pyrite, (210) face; phase contrast photomicrograph.

Unless specifically mentioned, all photomicrographs in this part are taken by a positive phase contrast microscope of the Zernike type.

neighboring steps are very narrow, individual contour lines cannot be seen even under high magnification. What one sees is a number of growth hillocks developing over the surface. In all these cases, crystal faces appear because of rapid two-dimensional spreading of growth layers parallel to the face, initiated either by a screw dislocation or two-dimensional nucleation. These faces may correspond to F faces according to the definition of Hartman and Perdok [41]. The observations to be presented in the following discussion are made on this type of crystal face.

Another type of crystal face is one which shows no growth layers but only straitions. Even if they are investigated under high magnification, these faces exhibit nothing but striations parallel to the neighboring faces. Most prismatic faces which appear in the prism zone of a prismatic crystal show such surface microtopographs, but other types of well-developed faces can also show similar characteristics. For example, the (210) faces of many natural pyrites, the scalenohedral faces of calcite, the prism faces of natural quartz, and the (211) faces of silicate garnets exhibit only striations, even though they develop faces as large or much larger than those on which visible contours develop freely. Although these faces can grow as large as the faces on which growth layers are seen, they cannot be regarded as F faces since they do not grow by a layer-by-layer growth mechanism. They should be treated as S faces. The so-called vicinal faces which occur as side faces of the polygonal growth pyramids developing on F faces have essentially the same characteristics as these faces. We shall exclude observations on these faces, since they show only straitions.

In Fig. 13A, B, and C, representative examples of the surface microtopographs of these faces are compared.

B. Methods of Observation and Measurement

Monomolecular growth spirals are not detectable with an ordinary optical microscope, except when spiral steps are heavily decorated by foreign materials (an example of which is shown in Fig. 14). In

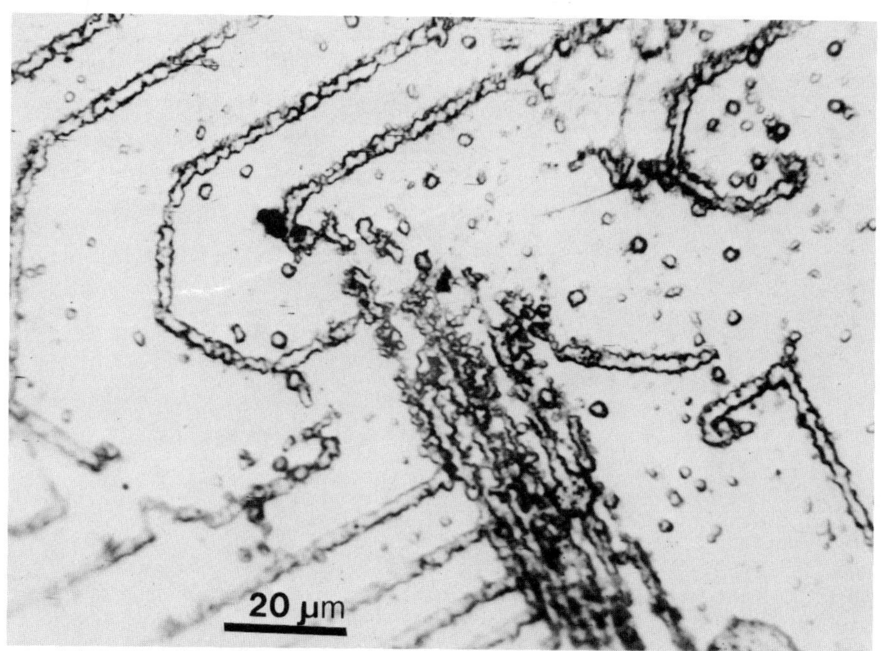

FIG. 14. An example of growth spirals whose steps are heavily decorated by foreign minerals. The height of the spiral layers is 10 Å. Unless they are heavily decorated, these are not detectable with an ordinary reflection optical microscope. Natural phlogopite (a kind of mica), (001); ordinary reflection photomicrograph.

general, either phase contrast microscopy of the Zernike type [61] or differential interference contrast microscopy of Nomarski or Yamamoto-Françon types [62] is required to detect growth layers thinner than about 100 Å.

In Zernike-type phase contrast microscopy, a high-contrast effect is secured by transforming a phase difference into an amplitude difference. This is realized by shifting the phase of a D wave by (1/4)λ and by absorbing the S wave considerably (see Fig. 15). The more the S wave is absorbed, the higher is the phase contrast effect obtainable. To observe spiral layers of less than 10 Å, a specially prepared phase contrast plate of high absorption (about 95%) is advisable.

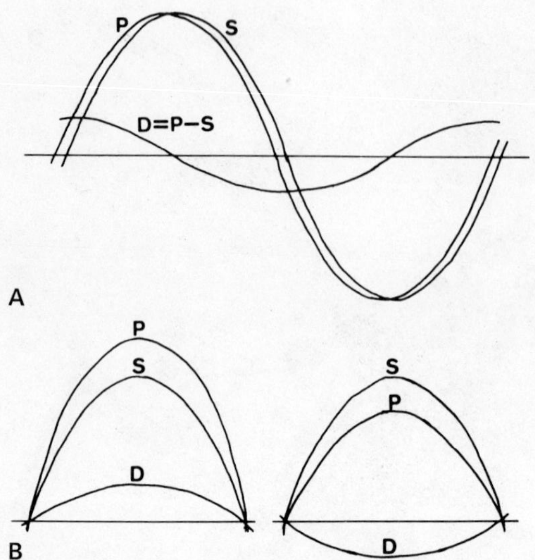

FIG. 15. The principle of phase contrast microscopy of the Zernike type: (A) the sine curves for the S, P, and D waves. The S wave corresponds to the wave of the medium, P to the particle in the medium, with the refractive indices of both waves being nearly equal. The D wave is obtained by subtracting S from P, and corresponds to the light beam diffracted at the boundary between the particle and the medium. Since there is no amplitude difference between S and P, the particle is not observable under the microscope. (B) If D is shifted by $\lambda/4$ in either direction, an amplitude difference is produced between S and P. This explains the principle of phase contrast microscopy, in which contrast (amplitude difference) is secured simply by shifting the D wave by an appropriate phase. Depending upon the direction of the shift, either positive or negative phase contrast is obtained. A higher contrast is obtainable when D becomes nearly equal to S, for which S should be absorbed considerably.

In phase contrast microscopy, thin steps are seen because a bright halo appears on one side of the steps and a dark band on the other side. A bright halo always appears on the higher side and a dark band on the lower side of a step with positive phase contrast, and vice versa for negative phase contrast (see Figs. 15 and 16 for actual examples). The topography, i.e., which side of a step is higher or lower, can therefore be unambiguously decided if a phase contrast microscope is used. This is almost impossible both in

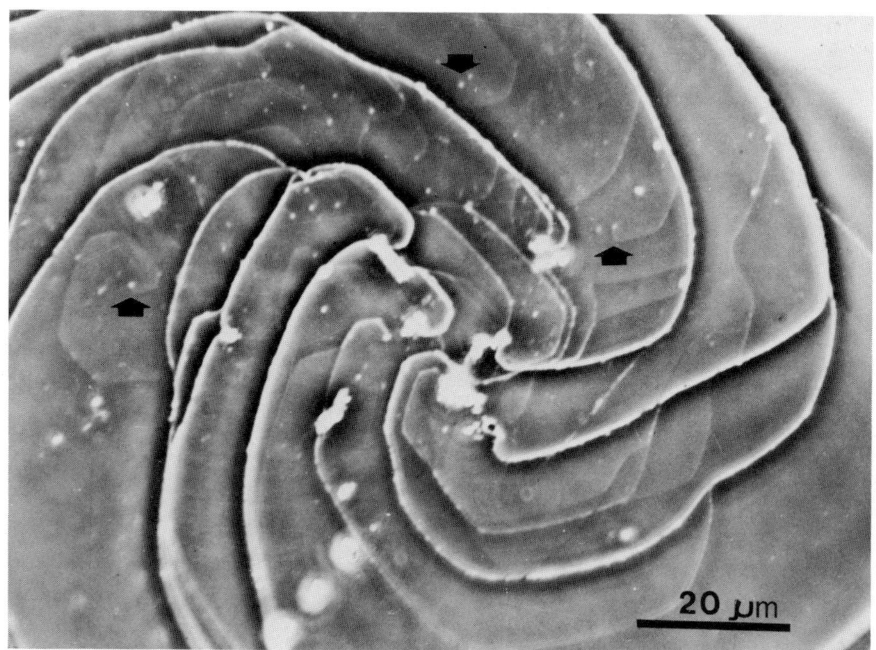

FIG. 16. Representative positive phase contrast photomicrograph. A bright halo appears on the higher side of a step. The nature of the halo varies depending on the step height. White dots are impurities selectively adsorbed at the points of emergence of both screw and edge dislocations. Spiral layers that originate from a group of screw dislocations and bunch together form thicker steps from the beginning. These exhibit wider and brighter halos than those originating from the single independent screw dislocations indicated by arrows. SiC, (0001).

ordinary reflection-type microscopy and in interference contrast microscopy. This advantage of Zernike-type phase contrast microscopy is, however, useful only for the steps of under $\lambda/2$ in height. A diffraction effect is introduced with steps higher than $\lambda/2$, resulting either in a diffused halo on both sides or in an interference color.

Another advantage of Zernike-type phase contrast microscopy is that one can roughly estimate step heights from the intensities and widths of the bright halo, since these increase nearly proportionally to the step heights (see Fig. 16).

There are, however, some disadvantages of phase contrast microscopy. Although it works perfectly well for a molecularly flat surface with sharp thin steps, a proper phase contrast effect is not obtainable for wavy or inclined surfaces. The best contrast is secured only for a surface which is vertical to the optical axis, and the effect sharply diminishes as the surface is inclined from this. This disadvantage may be overcome greatly by the use of differential interference contrast microscopy.

In Nomarski or Yamamoto-Francon interference contrast microscopy, a contrast effect is secured by splitting the light beam into ordinary and extraordinary waves by means of either a Savart plate (Y-F technique) or a Wollaston prism (N technique), and later combining the two waves by a polarizer (Fig. 17). Under the interference contrast microscope, a surface perpendicular to the optical axis of the microscope has an uniform interference color. The interference color on surfaces inclined to this depends on the orientation of the surface with respect to the polarizer. Complementary interference colors are seen on opposite sides of an elevation or depression (see Fig. 18). In effect, therefore, the topograph is seen as if it were illuminated by an oblique light. In this respect, interference contrast microscopy is more suitable for observation by the human eye,

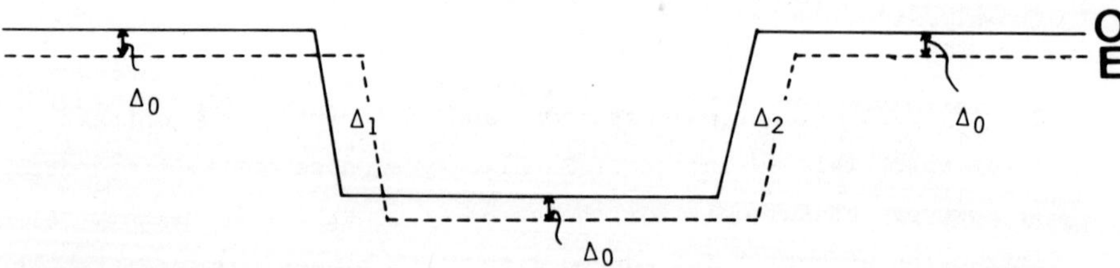

FIG. 17. Schematic explanation of interference contrast microscopy. An image is split by either a Wollaston prism or a Savart plate into ordinary (O) and extraordinary (E) rays. Interference between the two rays produces complementary colors for Δ_1 and Δ_2, with a background of a different interference color corresponding to Δ_0.

whereas phase contrast microscopy is best for photography. The interference microscope can also reveal steps of less than 10 Å. Both phase contrast and interference contrast microscopes are available commercially. In both types, in order to detect thin growth layers of less than a few tens Å, it is desirable to secure high reflectivity of the surface so that loss in phase contrast effect due to internal reflection or transmission is minimized. For this purpose, the surface should be silvered by vacuum evaporation, after thorough cleaning. Silvering the surface does not significantly modify the surface microtopography [63].

It should be stressed that even if all of the efforts described above are made, steps less than 10 Å high are not easy to detect. Experience and great patience are required before one may observe such thin growth layers under the microscope. The most important secret in observation of thin spiral layers is patience. Photographing the surface with special care to improve the contrast may improve visibility to a great extent. The 2.3 Å growth spirals observed on natural hematite [7] were not at all detectable under the most sensitive phase contrast microscope, but came out when they were photographed using the highest-contrast processing. It should be stressed again that even if one cannot see any feature on the surface under the microscope, this does not mean that growth spirals are absent. This is true even when phase contrast or interference contrast microscopy is used. Monomolecular spirals are usually found on extremely flat surfaces where no surface microtopographs are detectable even with these microscopes.

To measure step heights of such thin growth layers, the most sensitive and precise technique is the multiple-beam interferometry developed by Tolansky [6,64]. A step height of 15 Å can be directly measured by this technique. If an indirect measurement, such as dividing the total height of a spiral hillock by the number of steps consisting of the hillock, is applied by combination of phase contrast microscopy and multiple-beam interferometry, a step height of less than 3 Å can be measured [7].

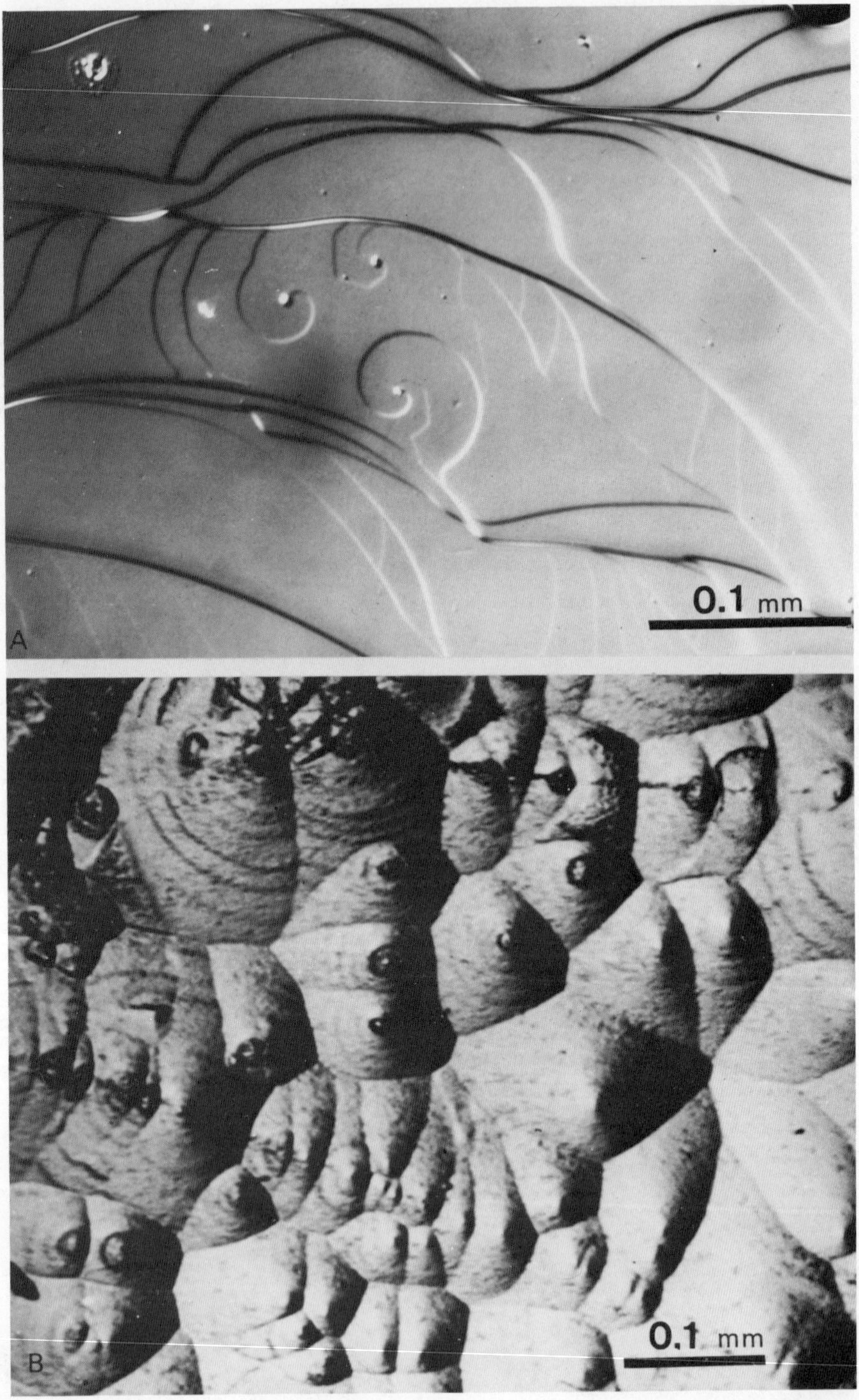

0.1 mm
A

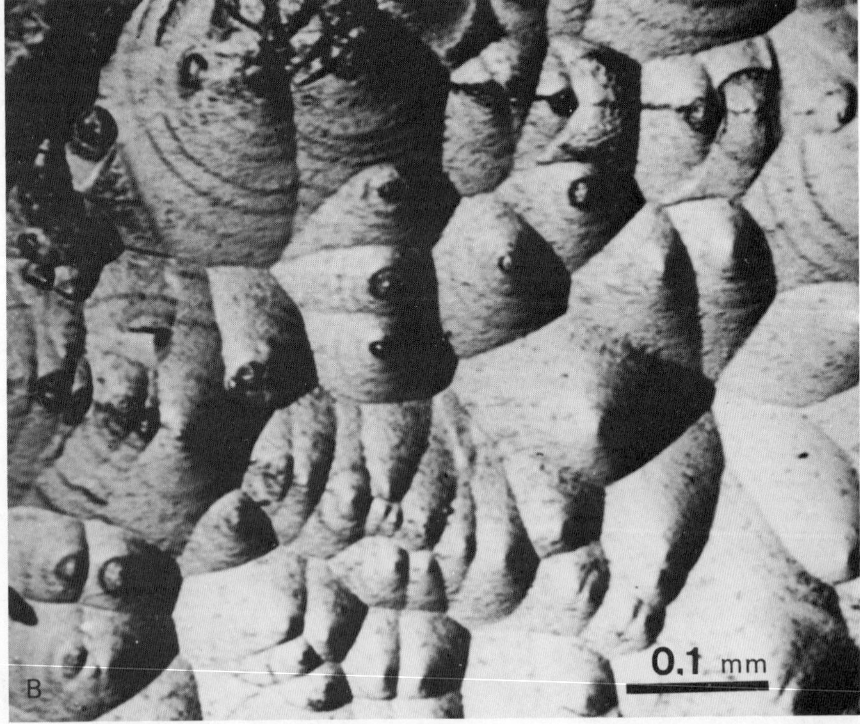
0.1 mm
B

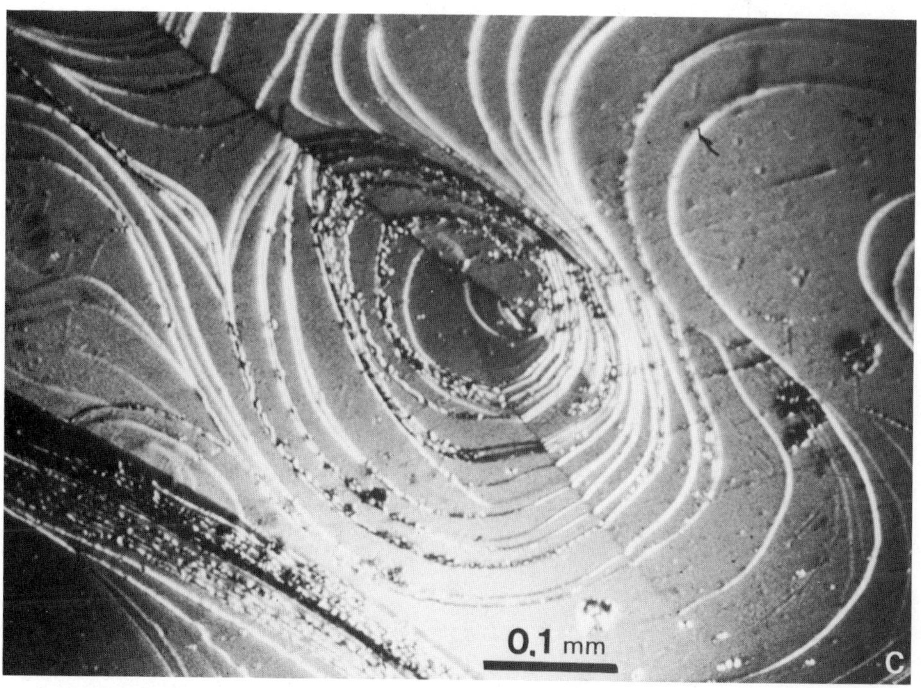

FIG. 18. Three examples of interference contrast photomicrographs: (A) Growth spirals with wide step separation. Spirals are elevated and have dislocation hollows at their centers (see Sec. III.D.1). Due to an optical illusion, the pattern may be seen as depressions, depending on the person, line of sight, etc. If the pattern appears to be a depression, look from the opposite side and then it will appear to be an elevation; SiC, (0001). (B) Growth hillocks with curved slopes, which actually consist of narrowly spaced steps; synthetic emerald, (0001) face. (C) A surface consisting of several inclined portions, each having a microtopograph similar to (A); natural hematite, (0001).

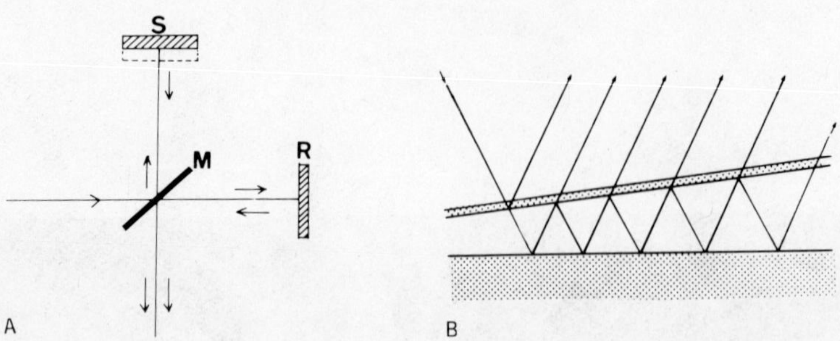

FIG. 19. A comparison of two-beam and multiple-beam interferometry: (A) Two-beam interferometry; S = specimen, R = reference mirror, M = half-mirror; only two beams contribute to the interference. (B) Multiple-beam interferometry; a large number of beams, reflected many times from the highly reflecting specimen surface, contribute to the interference.

In two-beam interferometry of the Michelson or Nomarski type, interference occurs between the two light waves, i.e., between the direct beam from the specimen surface and the beam reflected from the reference mirror (Fig. 19 A). The interference fringes in this type are as broad as 1/3 the fringe separation, which corresponds to about 1000 Å. Using such broad fringes, the limit of measurement is approximately 300 Å at best. If, however, a large number of beams is introduced, the interference fringes become much sharper as the number of beams increases (see Fig. 19 B). Then it is possible to diminish the fringe width down to as sharp as 1/50 the fringe separation ($\lambda/2$), which then corresponds to 40-50 Å. Multiple-beam interferometry thus enables direct measurement of step heights down to 15 Å. In Tolansky's multiple-beam interferometry, the surfaces of both specimen and the reference optical flat are silvered to secure high reflectivity so that the light beam can be multiply reflected between the two surfaces. In Fig. 20, two-beam and multiple-beam interferograms of similar surfaces are shown. Figure 21 shows an example of indirect measurements of steps thinner than 15 Å by the combination of phase contrast photomicrographs (a, c) and a multiple-beam interferogram (b). In Fig. 21 A, nine triangular pyramids are seen, whose side faces appear flat at lower magnification. Even when the side

FIG. 20. Comparison of two-beam and multiple-beam interferograms of similar surfaces: (A) two-beam interferogram; (B) multiple-beam interferogram. Both are interferograms of the (0001) surfaces of natural hematite, showing step patterns similar to Fig. 13 A.

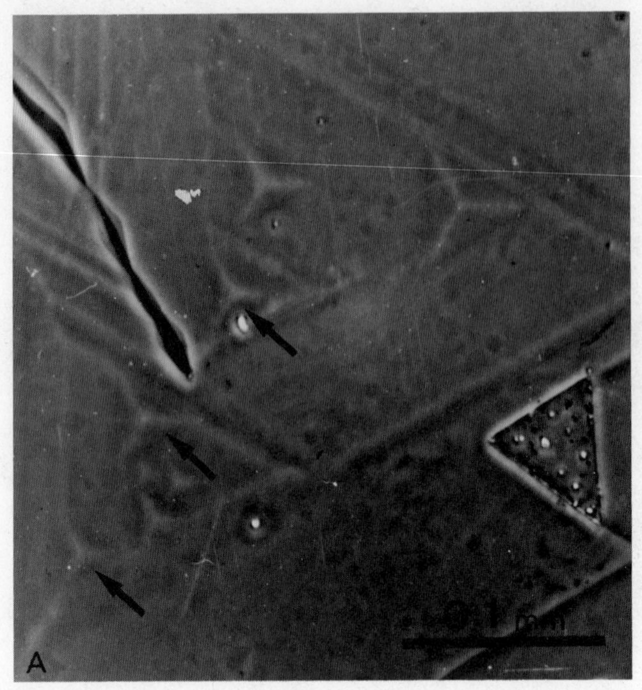
A
0.1 mm

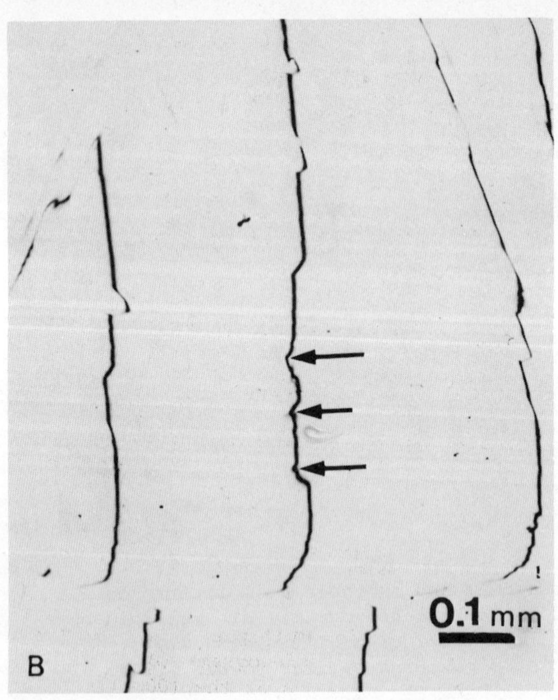
0.1 mm
B

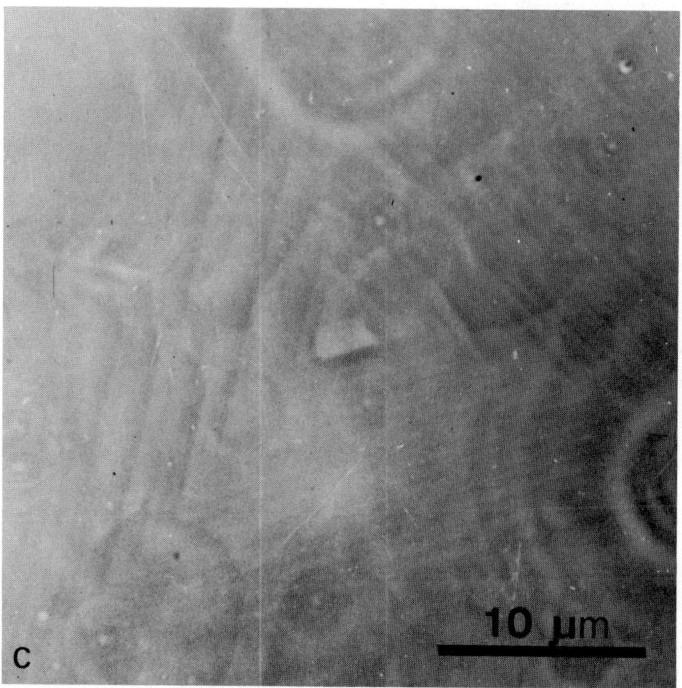

FIG. 21. Phase contrast photomicrograph (A), multiple-beam interferogram (B), and higher magnification phase contrast photomicrograph (C) of growth pyramids observed on the (0001) face of natural hematite from Ayumikotan, Hokkaido, Japan. By combination of two methods (measuring the heights of growth pyramids by multiple-beam interferometry and counting the number of steps consisting the pyramids on the phase contrast photomicrograph), the step height of single layers constituting the pyramids is measured to be 2.3 Å, which is the height of a single molecular layer of Fe_2O_3. Arrows in (A) and (B) correspond to the respective pyramids.

faces appear very flat under higher magnification, it is found that they actually consist of triangular growth spirals with equal step separation if special care is taken in photographing these pyramids under the highest magnification (see Fig. 21C). The number of the spiral layers constituting the pyramids can be counted on the photograph. The fringe crossing the summits of the pyramids shows bends as indicated by arrows in Fig. 21 B on the interferogram (arrows in Fig. 21B correspond to arrows in Fig. 21A), from which total heights of the pyramids are measured. These values range from about 20 to 120 Å. Using both values, the step height of single spiral layer was determined to be 2.3 Å, corresponding to the height of a single Fe_2O_3 layer.

For electron-microscopically small crystals, the optical methods described above are not applicable. For the study of surface microtopographs of such crystals, the decoration technique of electron microscopy may be useful. This technique has been developed and applied extensively to study growth and dissolution of NaCl crystals by the group of Bethge [8-10]. It was also successfully applied to observe surface microtopographs of some clay minerals of different origins by several workers [11,12]. Growth spirals of monomolecular height were revealed on the (001) faces of clay minerals.

In this technique, specimens are heated to an appropriate temperature, so that the surface is cleaned and the mobility of evaporated gold is increased. Then gold is flash-evaporated, followed by carbon coating. After these procedures the sample is taken out of the evaporation unit and dissolved in an appropriate solvent, so that only the surface replica film is obtained. Figure 22 is an example which clearly shows the surface microtopography of a clay mineral. The step height measurement can be made in this technique by dividing the height of a bunched layer, which can be measured by means of shadowing technique, by the number of single layers constituting the bunched layer.

FIG. 22. Electron photomicrograph of the (001) surface of a dickite crystal, showing growth spirals with a step height of 10 Å. Taken by the decoration technique of electron microscopy. Photo by Y. Koshino.

C. Ideal Growth Spirals

1. *Archimedean Spirals*

We at first focus our attention on the morphology of growth spirals originating from isolated single screw dislocations. Interaction and cooperation among spiral layers originating from different screw dislocations are discussed in Sec. III.E.

According to BCF [1], Cabrera and Levine [44], Ohara and Reid [45], and Müller-Krumbhaar et al. [46], when a spiral originating from an isolated single screw dislocation is formed under a steady growth condition and is not perturbed by spiral layers from other sources, it takes an Archimedean form, i.e., the spiral has an equidistant step separation λ_0 after the first turn. Figure 23 shows a computer-calculated ideal Archimedean spiral. Here two sets are shown,

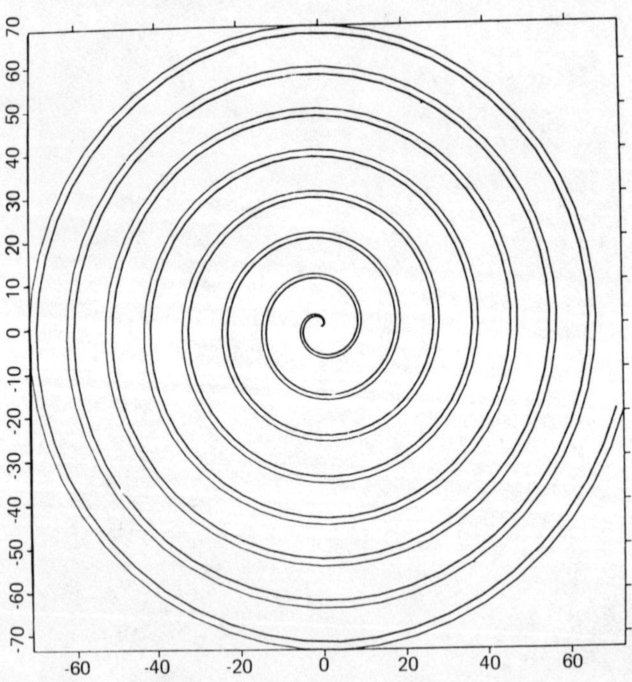

FIG. 23. Computer-calculated Archimedean spiral. Outer step corresponds to $19r^*$, inner step to $4\pi r^*$.

the outer one corresponding to $19r^*$, the inner one to $4\pi r^*$. The step separation of the first turn is slightly different from the separation farther from the center due to the overlap of diffusion fields at the center, and increases during the first few turns. Apart from this, λ_0 is constant. If an actual spiral deviates from this, either a perturbation in growth conditions or an interaction with neighboring spirals should be suspected.

On actual crystals, ideal growth spirals are not commonly observed, but occasionally we meet spirals which can be regarded as almost ideal Archimedean spirals. Figure 24A and B demonstrate such spirals on the (0001) face of SiC crystals synthesized by the Lely and Acheson methods, respectively [65]. Two representative spirals of circular and polygonal morphology are shown. In general, the area covered by such an Archimedean spiral originating from a single screw dislocation is very small, less than 1 mm^2. It is very rare that the entire surface of a face is covered by such an ideal spiral originating from a single screw dislocation. When crystals are synthesized under very carefully controlled conditions, the area covered by a single spiral can be as wide as 10 mm^2, if perturbations are not present. Such an example is shown in Fig. 25 for a SiC crystal synthesized by the Lely method. In most crystals, especially natural crystals, such a wide coverage is usually not seen. Commonly, a large number of spirals, either composite or independent, occur on one surface. This is principally due to the high probability of coalescence with other individuals while a crystal is growing, and also to the movement and concentration of screw dislocations into a small area during growth, as exemplified by Sunagawa on natural hematite crystals [66]. The point of coalescence provides new sites from which growth layers originate, forming composite spirals. So far, the maximum area observed to be covered by a spiral originating from a single dislocation on a natural crystal is 1 mm^2 for the case of a hematite (0001) face. With tiny crystals like clay minerals, the surface of crystal faces smaller than 1 μm across is often covered by a growth spiral originating from a single screw dislocation. An example is shown in Fig. 22. However, on wider surfaces more than one spiral center appears [12].

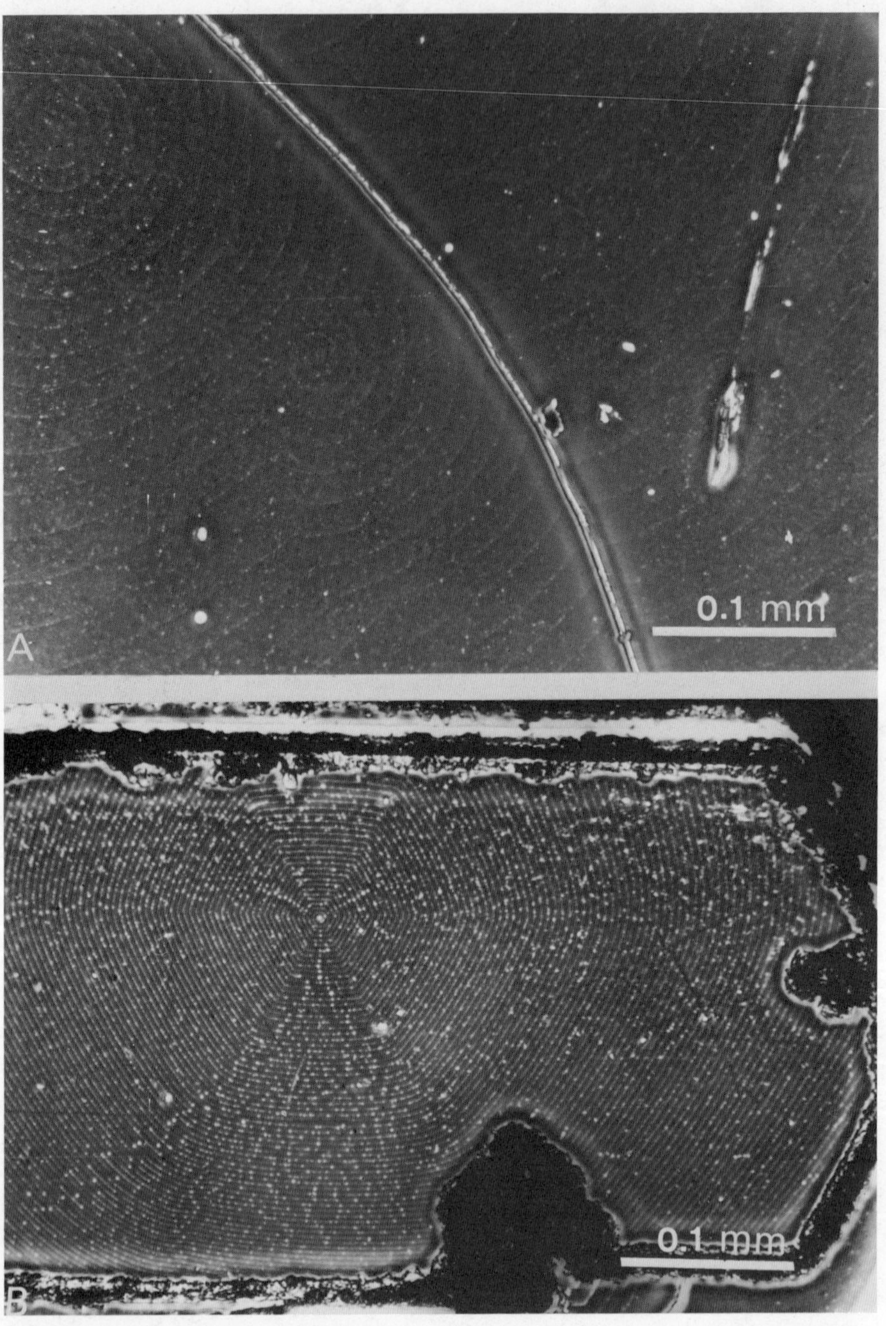

FIG. 24. Examples of circular (A) and polygonal (B) Archimedean type spirals observed on (0001), SiC.

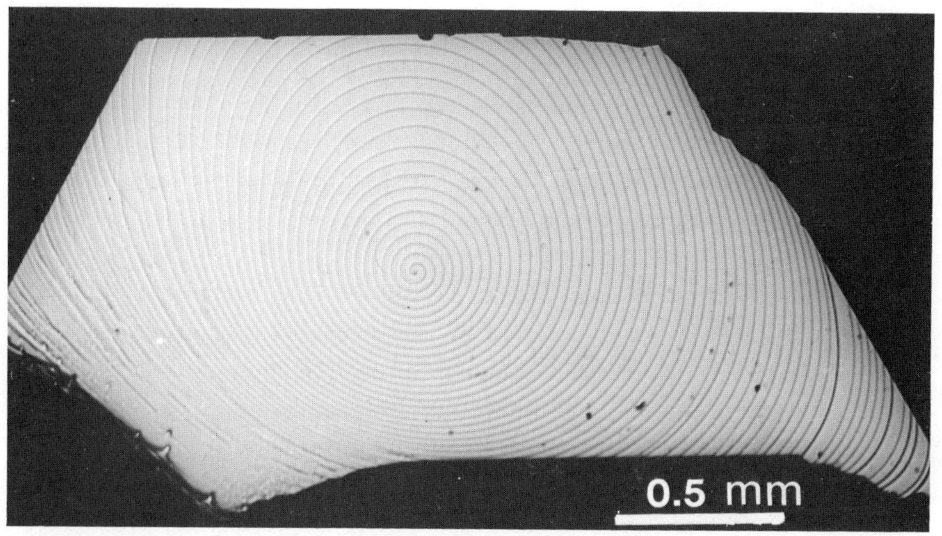

FIG. 25. An example of a face entirely covered by a growth spiral originating from a single dislocation. Eccentricity in the spiral pattern is due to supersaturation gradient over the surface (see Sec. III.D.2); (0001), SiC. Photo. by S. Itabashi.

The step height of spiral layers originating from a single screw dislocation is defined by the Burgers' vector of the dislocation. It is most commonly 1 unit cell in height, but can be a fraction of a unit cell or a small multiple. In the case of natural hematite, growth spirals of 2.3, 4.6, 7, 14, and 21 Å have been observed and measured [7]. This corresponds to 1/6 to 3/2 unit cell height in a hexagonal cell (c_0 = 13.73 Å in the hexagonal cell; 4.6 Å in the rhombohedral cell: the height of Fe_2O_3 single layer is 2.3 Å). In the case of SiC, which is a polytypic crystal, ideal growth spirals with step heights of a few hundred Å have been observed [4,5,67]. However, such spirals are uncommon, since dislocations having big Burgers' vectors are energetically less stable and can easily decompose into many dislocations of smaller Burgers' vectors.

Although the computer simulations predict a step roughness which should be detectable by optical or electron microscopy (see Sec. II.B.6), observed growth spirals usually have very smooth steps, irrespective of their morphology, circular or polygonal. Step roughness (denticulation) has not been observed within the limit of lateral resolution of optical microscopy, and even of electron microscopy, except for a special case to be discussed later. When foreign particles are present, steps can be denticulated, but such denticulation is essentially different from the step roughness expected from the computer simulations and can easily be distinguished from the latter by the occurrence of foreign particles along the growth steps (Fig. 26).

The surfaces of ideal spiral layers are atomically flat, so no detectable roughness is observed by phase contrast or interference contrast microscopy or multiple-beam interferometry.

It should also be noted that the steps of ideal growth spirals form very sharp cliffs in general. Thus no detectable slope is seen on the multiple-beam interferograms. Although it is impossible to measure the inclination of the steps (there is no means of measuring the inclination of such thin steps), it is reasonable to assume that the steps of ideal growth spirals have the characteristics of sharp cliffs rather than of gentle slopes. Of course this is not so for higher steps or bunched thick steps.

Since schematic drawings of spiral growth often give the mistaken impression that the ratio of the step separation to the step height is on the order of 10 or even less, it should be stressed here that this is not actually so. Although the step separation λ_0 can vary considerably depending on growth conditions, especially on supersaturation, the ratio of λ_0 to step height h generally ranges from 10^2 to 10^3 for aqueous solution growth, to 10^3 to 10^4 or even larger for vapor growth [15,68]. The real image of growth spirals is therefore analogous to travel of 100 to 10,000 m on an extremely flat plane and then meeting a sharp cliff of 1 meter in height. One should not misinterpret the real profile of growth spirals from the figures in textbooks.

Judging from the characteristics of ideal growth spirals described above, one may assume that they are formed by a typical spiral growth

FIG. 26. Triangular growth spiral with denticulated steps due to adhesion of foreign particles during growth; (0001), SiC.

mechanism through the incorporation of atomic, ionic, or molecular entities. Such spirals have been observed most commonly on crystals grown from the vapor [15,68]. In fact most vapor-grown crystals with well-developed crystal faces typically exhibit such spirals. These have also been observed occasionally on crystals grown from high-temperature solutions (e.g., flux growth) [69], as well as from aqueous solutions with a low supersaturation [12]. However, on most crystals grown from an aqueous solution, such as NaCl, KCl, and natural mineral crystals of hydrothermal origins, ideal growth spirals have rarely been observed. They usually exhibit growth layers with macrostep heights which are not due to bunching of thinner monomolecular layers, denticulated steps, or wavy surfaces. These characteristics are essentially different from those of the ideal growth spirals described above. It is difficult to attribute these characteristics to mechanisms similar to those discussed above. Unless much larger growth units are assumed to be incorporated in the crystal, these characteristics cannot be accounted for. As a typical example, growth spirals observed on the (001) faces of synthetic diamonds are shown in Fig. 27. We shall discuss these macrostep spirals at the end of Sec. III.

2. *Step Separation λ_0: Actual Profile of Growth Spirals*

As discussed in Sec. II, when a spiral is formed under steady-state growth conditions, it takes an Archimedean form after a few central turns, i.e., λ_0 is uniform apart from a somewhat wider step separation at the central few turns. Thus, a straight lateral profile is expected for an ideal growth spiral apart from the slightly less inclined center, provided that surface diffusion and the back stress effect are not taken into account. However, such ideal Archimedean-type spiral profiles can normally be observed only in small areas. An actual phase contrast photomicrograph and a multiple-beam interferogram showing the profile of an almost Archimedean spiral covering a wide area is illustrated in Fig. 28. In most cases, the coverage is much narrower, and the spiral profiles usually deviate from the ideal profile; namely λ_0 becomes either wider as spiral layers go outerward or narrower close to the edges. There are also cases in

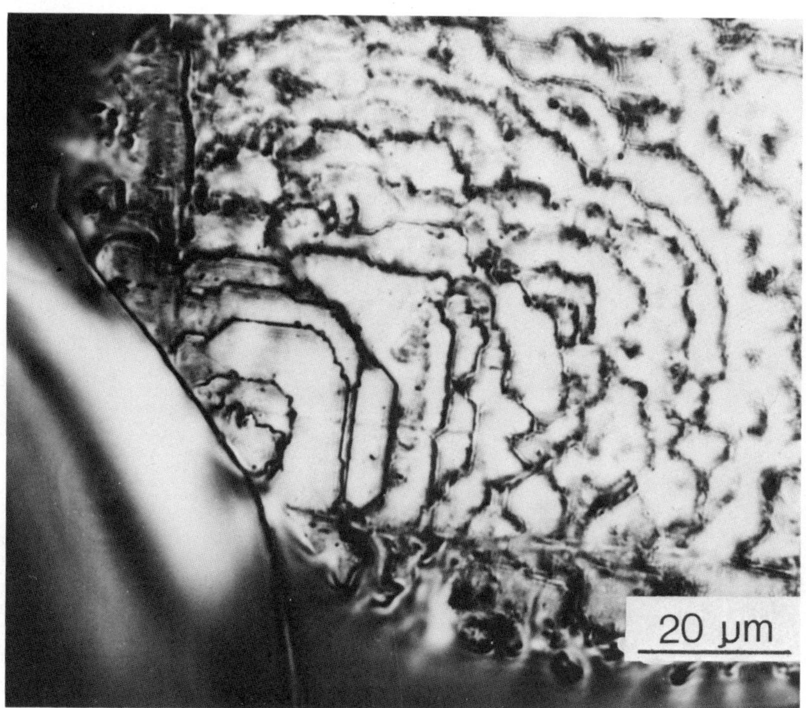

FIG. 27. An example of growth spirals commonly observed on crystals grown from high temperature or aqueous solutions; (001), synthetic diamonds. Step height is about 1000 Å. Note denticulated steps and wavy surfaces.

which λ_0 at the central portion of a spiral beocmes either much narrower or wider compared to λ_0 far from the center. These deviations from the ideal morphology are due to perturbations in growth conditions, as discussed below.

a. λ_0 *NEAR THE CENTER OF A SPIRAL*

The simplest case is deviation from the Archimedean-type step separation at the spiral center. Even when a spiral shows almost equidistant step separation far from the center, it is often observed that at the spiral center the separation becomes either much narrower or much wider than the separation far from the center. Two examples are shown in Fig. 29 A and B. On synthetic crystals, narrower step

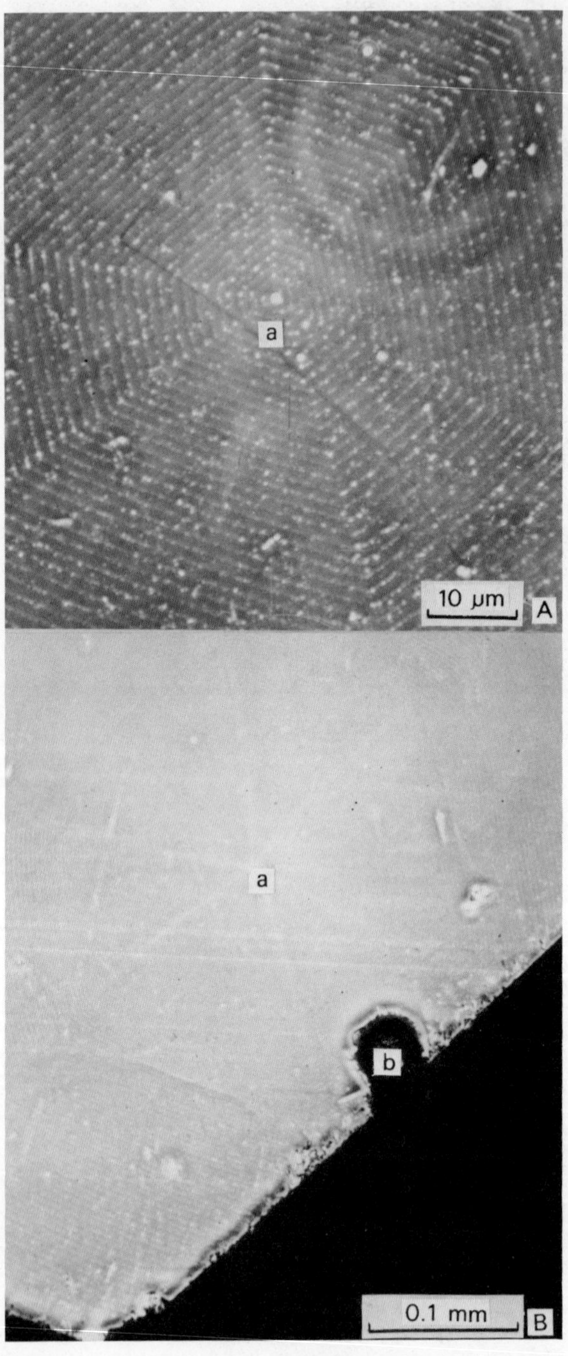
a
10 μm
A
a
b
0.1 mm
B

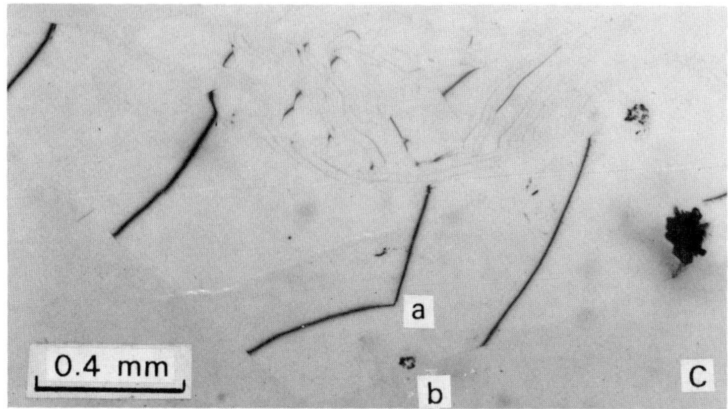

FIG. 28. Ideal Archimedean-type spiral covering a wide area and its profile: (A) higher magnification of the spiral center a; (B) lower-magnification photomicrograph, showing the position of the spiral center a and the crystal edge b; and (C) multiple-beam interferogram, whose central fringe crosses the spiral center A, from which the profile of the spiral can be estimated. The step height of a spiral layer is 15 Å.

separations at the center are commonly observed, an example being shown in Fig. 29 A. This is attributed to the rapid supersaturation increase at the final stage of growth due to the rapid temperature drop. On natural crystals, wider step separations at the spiral center are more commonly observed, as shown in Fig. 29 B. The latter type is often observed on spirals having an almost Archimedean form with a wide step separation, more than 10^4 in the ratio of λ_0/h, i.e., crystals grown with a very low supersaturation. A gradual diminishing of supersaturation may account for this profile.

b. λ_0 *APART FROM THE CENTER*

It is almost exceptional that constant λ_0 is maintained for more than 100 turns. There can be many perturbations causing this. Since a detailed discussion of the origin of such perturbations will be given in Sec. D, only possible causes are mentioned now:

1. Perturbation near thick bunched layers
2. Perburbation near the edge of a crystal (morphological effect)

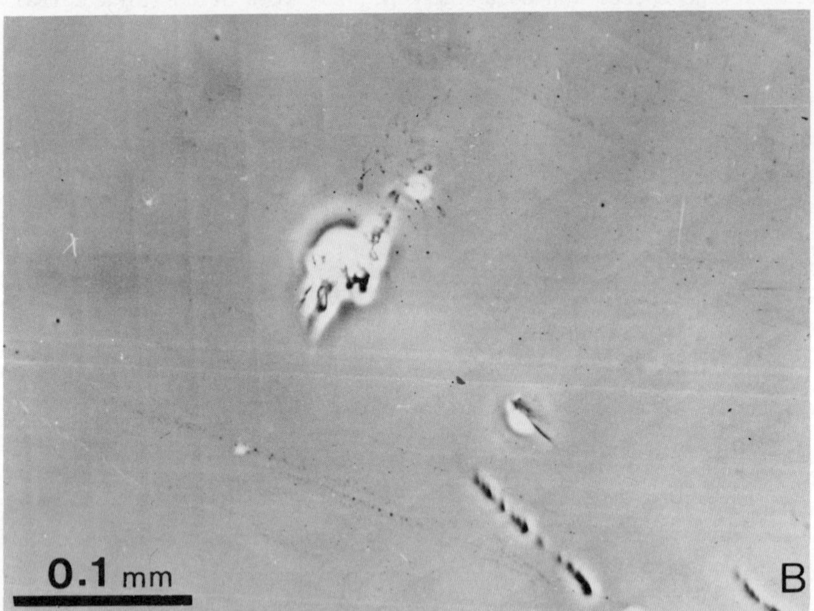

FIG. 29. Examples to show abrupt decrease (A) and increase (B) of step separations at the spiral centers: (A) (0001), SiC; note also interlacing pattern (see Sec. III.C.4) and impurity decoration of the steps; (B) spiral with step height of 4.6 Å; (0001), natural hematite.

3. Perturbation due to re-entrant corner formed where two spiral layers meet
4. Perturbation due to supersaturation gradient over the surface
5. Perturbation due to Berg effect [81] near the edge of a crystal

3. Relation between λ_0 and h

Although neither the BCF theory [1] nor the recent computer calculations consider the effect of difference in step heights on step separation, observations clearly demonstrate that λ_0 depends strongly on h. Naturally this effect should be compared among spirals having different step heights which are formed simultaneously under the same growth conditions. Figure 30 gives an impressive example showing the dependence of λ_0 on h. Three groups of spirals are seen on this surface. Judging from the mutual relation among the three, it is safe to assume that they appeared almost simultaneously, i.e., under the same growth condition. The spiral in a has a step height of 350 Å, that in b 250 Å, and in c 60 Å. The λ_0 of a is 15 μm, b 14 μm, and c 7 μm. A clear dependence of λ_0 on h is noticed; the thinner the spiral step, the narrower is the step separation. However, the increase of λ_0 as h increases is not linear. A four-fold change in h results in only a two-fold change in λ_0. Since $\lambda_0 = 19\gamma/\Delta\mu\cdot\Omega$ (see Sec. II.B.1), this means that an increase in step height does not correspond to a proportional increase of λ_0. This is discussed in Sec. II.B.7.

4. Polygonization

The morphology of an ideal spiral is defined by two factors: (1) whether it is circular or polygonal; and (2) step separations between the successive arms.

Whether a spiral takes a circular or polygonal form is principally controlled by the roughness of the spiral step. If the step is rough, i.e., the kink density is high, the step can advance independently of crystallographic direction, and the spiral becomes circular. If the step is smooth, i.e., the kink density is low, the crystallographic directions show up more and the spiral becomes more

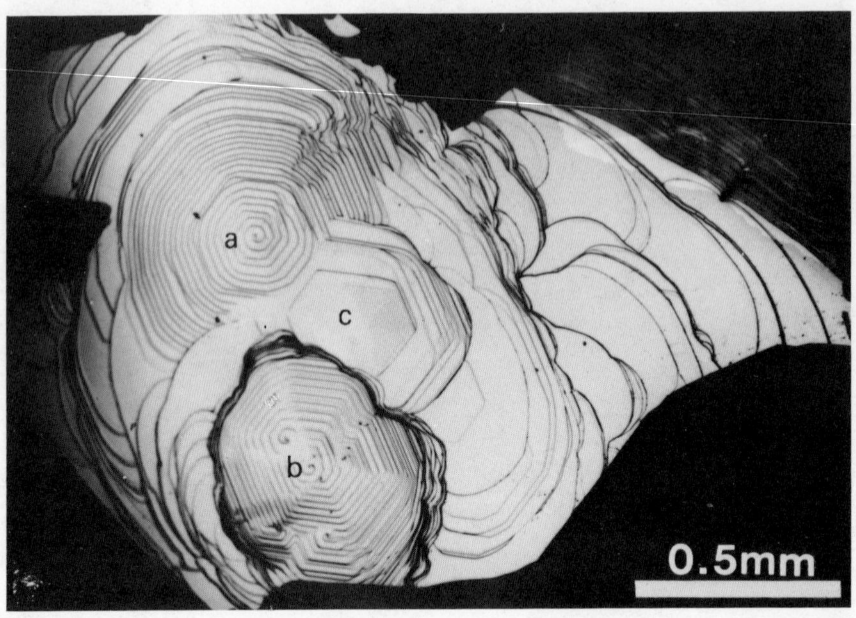

FIG. 30. Three growth spirals a, b, and c on the same surface, each having different step height. Note the difference in step separations. Spirals a and b have much higher step heights and wider step separations than the spiral c. Photo. by A. Kohuchi.

polygonized. This is analogous to the concept applied to the roughness of solid-liquid interfaces; thus an idea similar to that introduced by Jackson [37] for the latter is applicable.

As discussed in Sec. II, similarly to Jackson's α factor, the higher the α value, the less is the kink density of a step, or the smoother is the step. In other words, the larger the α value, the more polygonized is the spiral. However, the critical α value for the transition from circular to polygonal is larger than that for the solid-liquid interface transition from rough to faceted. According to Jackson [37], the solid-liquid interface transforms from rough to smooth at $\alpha > 3$. According to computer simulation for the (001) face of a Kossel crystal by Müller-Krumbhaar et al. [46], polygonization apart from the center of a spiral occurs in the range of $7 < \alpha < 8$, and at the center at $\alpha > 25$. In Figs. 4, 5, and 6, computer-calculated spirals are shown for different α values. For actual crystals with more complicated structures, polygonization may occur at smaller values than these.

As can be seen from Figs. 4, 5, and 6 polygonization apart from the center occurs at much lower α values than in the center. It follows from this that even if a spiral is polygonized, the spiral step near the origin can have a circular form. If polygonization occurs at the center, this suggests growth conditions corresponding to very high α values. In Fig. 31 A and B, both cases are shown. Figure 31A is observed on a hexagonal spiral of SiC, whereas Fig. 31 is on a triangular spiral of natural hematite. It is seen as a general tendency that most growth spirals observed on natural crystals

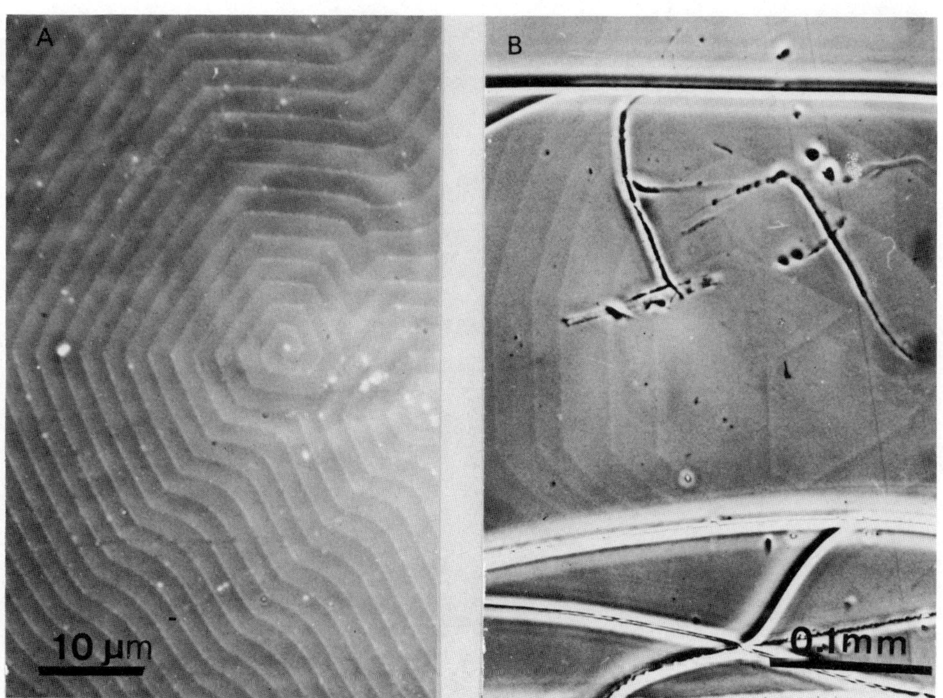

FIG. 31. Circular (A) and polygonal (B) centers on polygonized spiral: (A) (0001), SiC, 15 Å; (B) (0001), natural hematite, 7 Å.

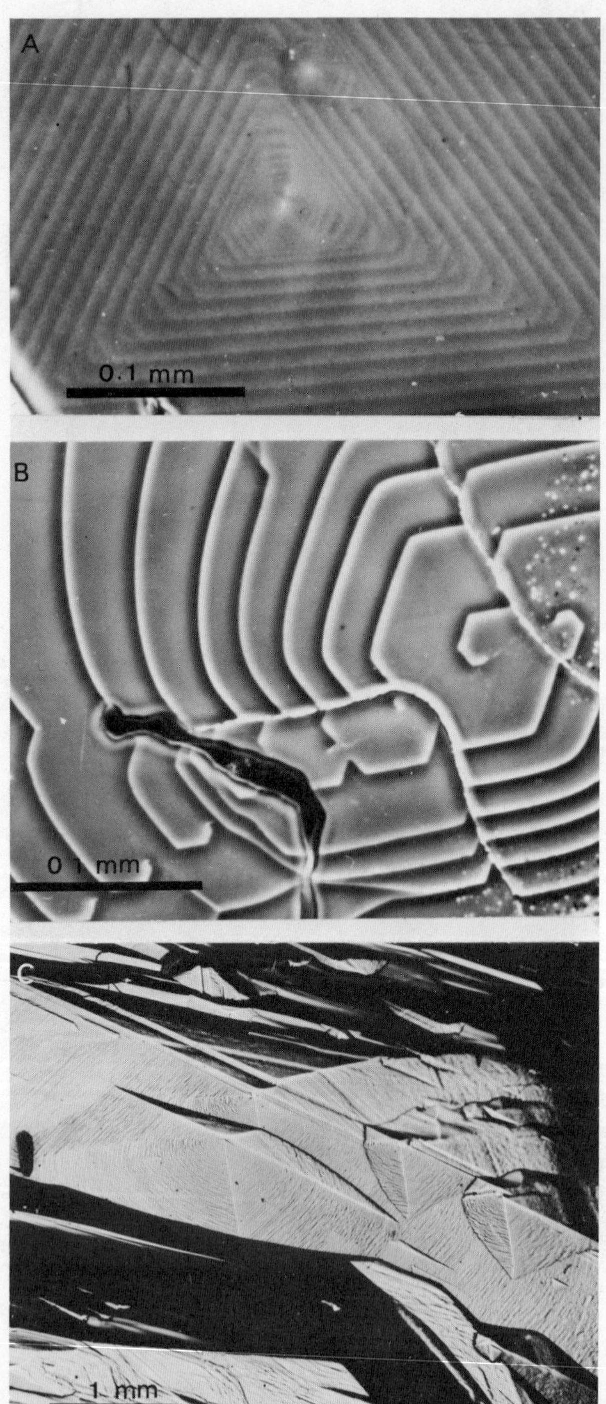
A
0.1 mm
B
0.1 mm
C
1 mm

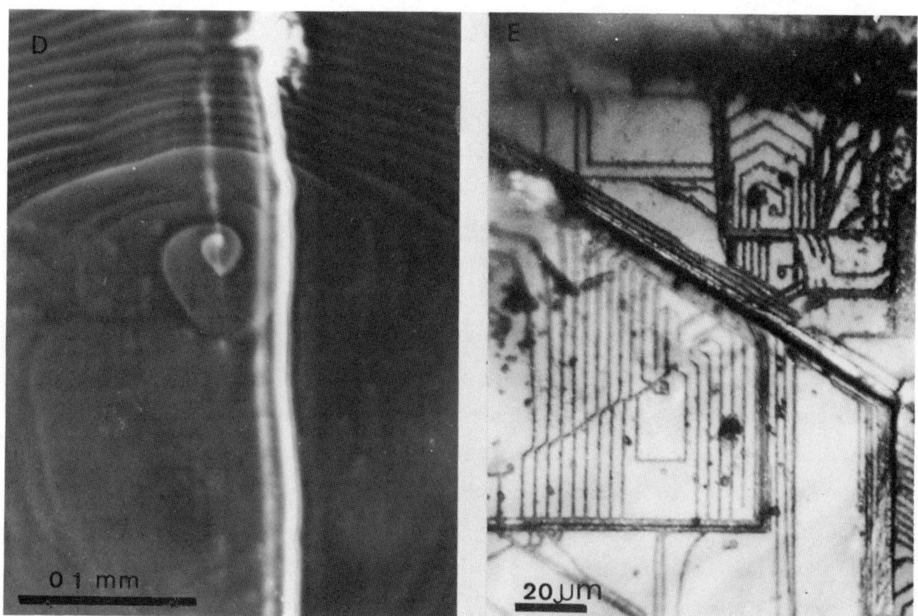

FIG. 32. A few examples of polygonal spirals showing the relation between the morphology of spirals and the symmetry of the face: (A) (0001), SiC; rhombohedral polytype; (B) (0001), SiC; hexagonal polytype; (C) $(10\bar{1}0)$, synthetic quartz; eccentric growth pyramids due to enantiomorphism of the crystal; (D) $(10\bar{1}1)$, hematite; symmetry element m; (E) (001), phlogopite, monoclinic; this is a twinned crystal, with composition plane (001); note opposite orientations of the spirals.

show polygonization at the center, whereas most synthetic crystals do not, even if spirals are as a whole polygonized.

When a spiral is polygonized, the morphology is defined by the stronger bonds in the face, i.e., periodic bond chain [41], since steps along the stronger bonds contain a lower density of kinks than those along the weaker bonds. For the (001) face of a Kossel-like crystal, [01] directions are the strongest PBC directions. Thus the spiral on this face is a square bounded by [01] directions. Under the conditions corresponding to lower α values, the square becomes truncated by [11] directions, the next strongest directions, while for still lower α the corners become rounded off. It follows from this that the symmetry of polygonized spirals should be in accordance with the symmetry of the face, which is, in fact, universally observed [70]. In Fig. 32 a few examples demonstrating this relation are shown. As can be seen in this figure, the (0001) face of rhombohedral crystals shows triangular spirals (SiC), of hexagonal crystals (SiC) hexagonal spirals; the $(10\bar{1}0)$ face of enantiomorphic crystals shows eccentric growth pyramids (synthetic quartz), whereas the $(10\bar{1}1)$ face of rhombohedral crystals (hematite) and the (001) face of monoclinic crystals (phlogopite) show a spiral with only one symmetry plane.

Figure 33 shows growth spirals observed on the (0001), $(10\bar{1}0)$, and $(21\bar{3}1)$ faces of a synthetic emerald grown by the hydrothermal technique. The (0001) face exhibits hexagonal spirals with sixfold symmetry, whereas the $(10\bar{1}0)$ face has spirals with a rectangular form (two symmetry planes), both in accordance with the symmetry of the respective faces. The $(21\bar{3}1)$ face on the same crystal exhibit circular morphology of growth pyramids. It has also been observed that opposite faces of a polar crystal like ZnS [71] or SiC [67], and also faces on enantiomorphic crystals like quartz [72], show quite different surface microtopographs.

Figure 34 shows a growth spiral observed on the basal plane of a magnetoplumbite crystal. This crystal belongs to the hexagonal system, but the unit cell consists of two equivalent blocks rotated 180°. Two spiral layers originate from one single screw dislocation, each having the height of a half unit cell. As can be clearly seen, each spiral layer shows a triangular morphology and the spiral as a whole shows a hexagonal morphology. Due to the difference of advancing rates between the upper and the lower triangular spiral layers in the same direction, an interlaced pattern appears along the six corners of the hexagonal pattern, where the two layers have an intermediate advancing rate (Fig. 35). Such interlacing patterns can appear on any polytypic crystals in which zigzag stackings exist in the structures [73,74].

These observations clearly demonstrate that when ideal monomolecular growth spirals are polygonized, their symmetry follows the symmetry of the face, not of the whole crystal, and very well reflects the structural characteristics.

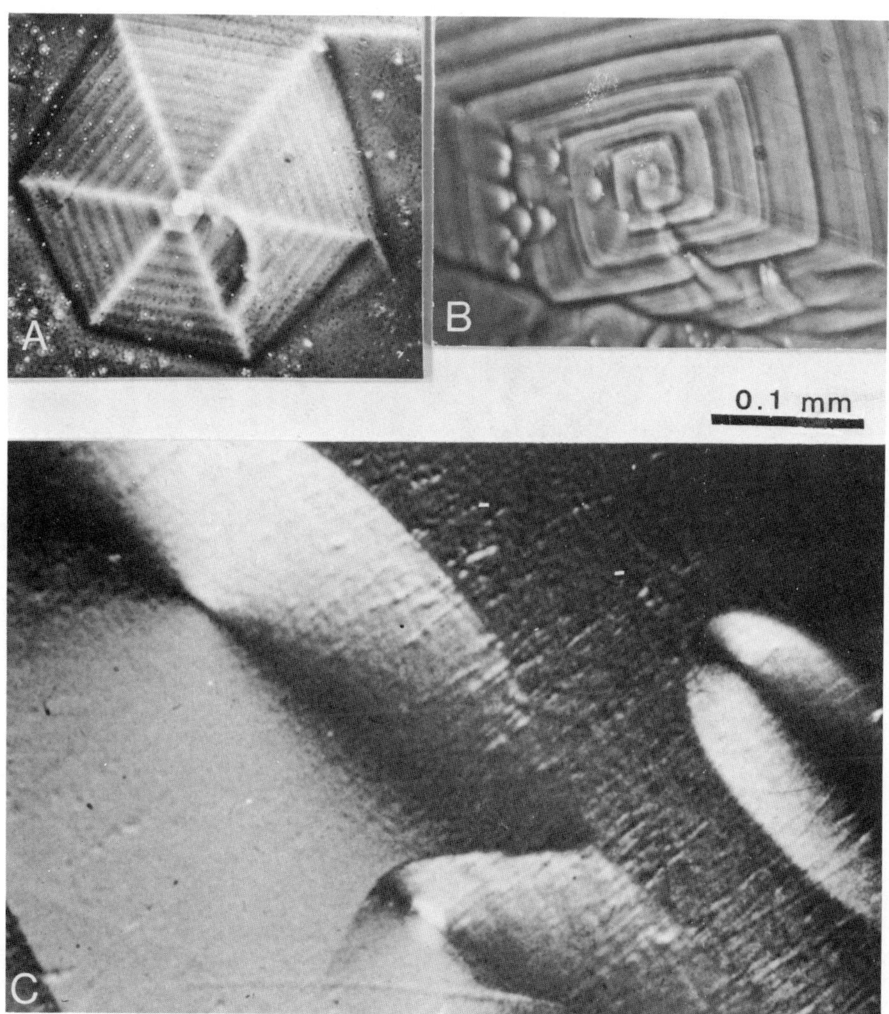

FIG. 33. Growth spirals observed on (A) (0001); (B) (10$\bar{1}$0); and (C) (21$\bar{3}$1) faces of the same emerald crystal synthesized by hydrothermal technique. Note the difference in symmetry elements and in the degree of polygonization.

Since α also depends on the orientations through $\xi = E^{sl}/E^{cr}$ (see Sec. II.A.4), different crystal faces on the same crystal can

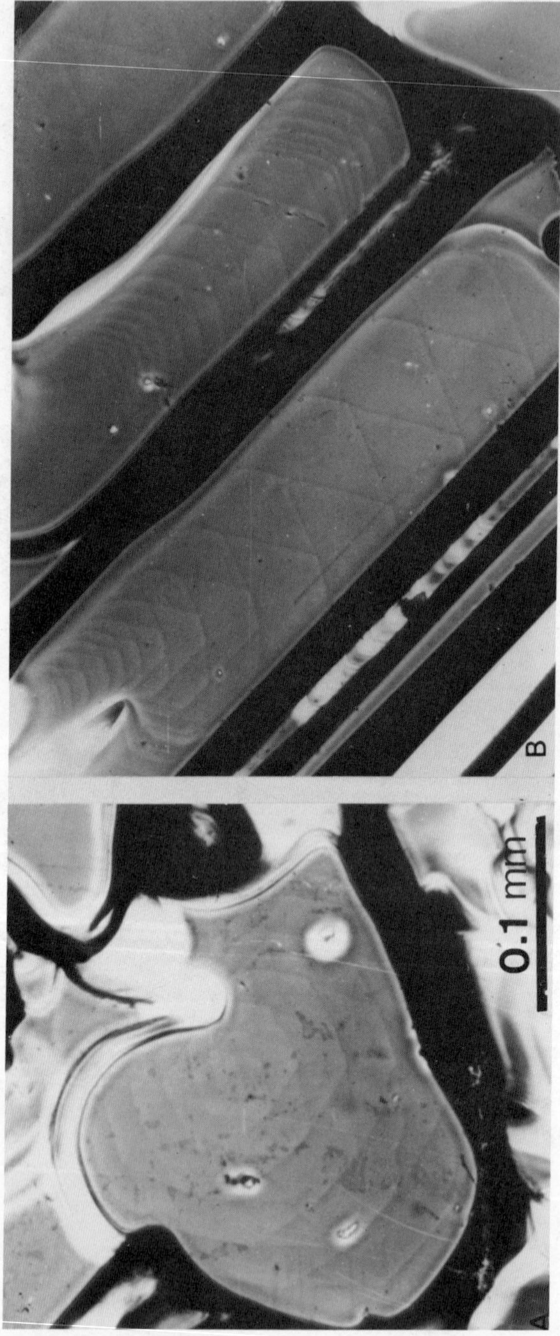

FIG. 34. Interlaced spiral observed on the (0001) face of a flux-grown magnetoplumbite: (A) spiral center; (B) far from the center.

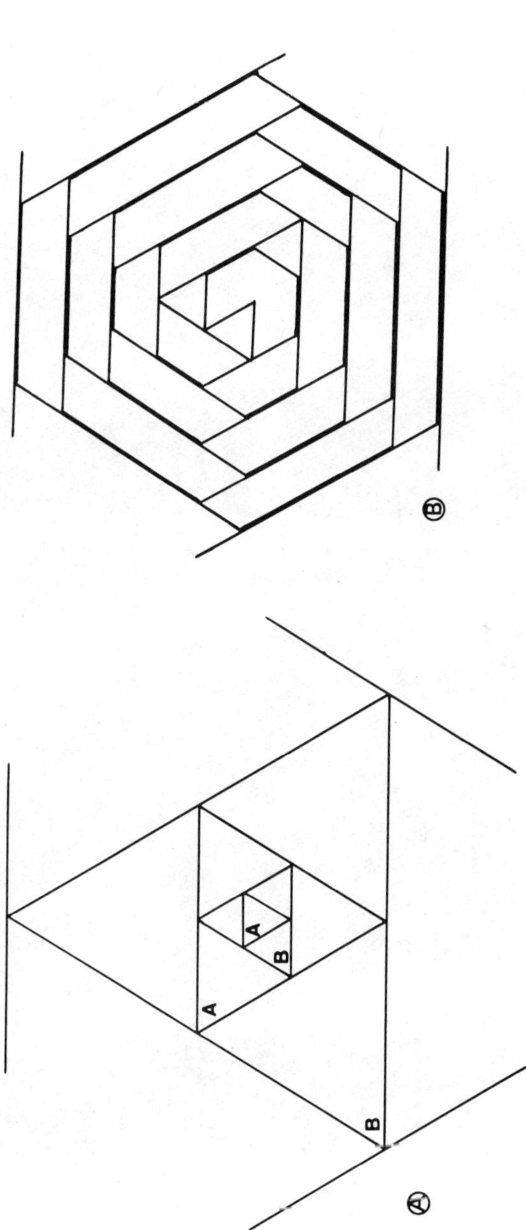

FIG. 35. Illustration of how an interlaced pattern is formed by two spiral layers of opposite orientations originating from a single dislocation.

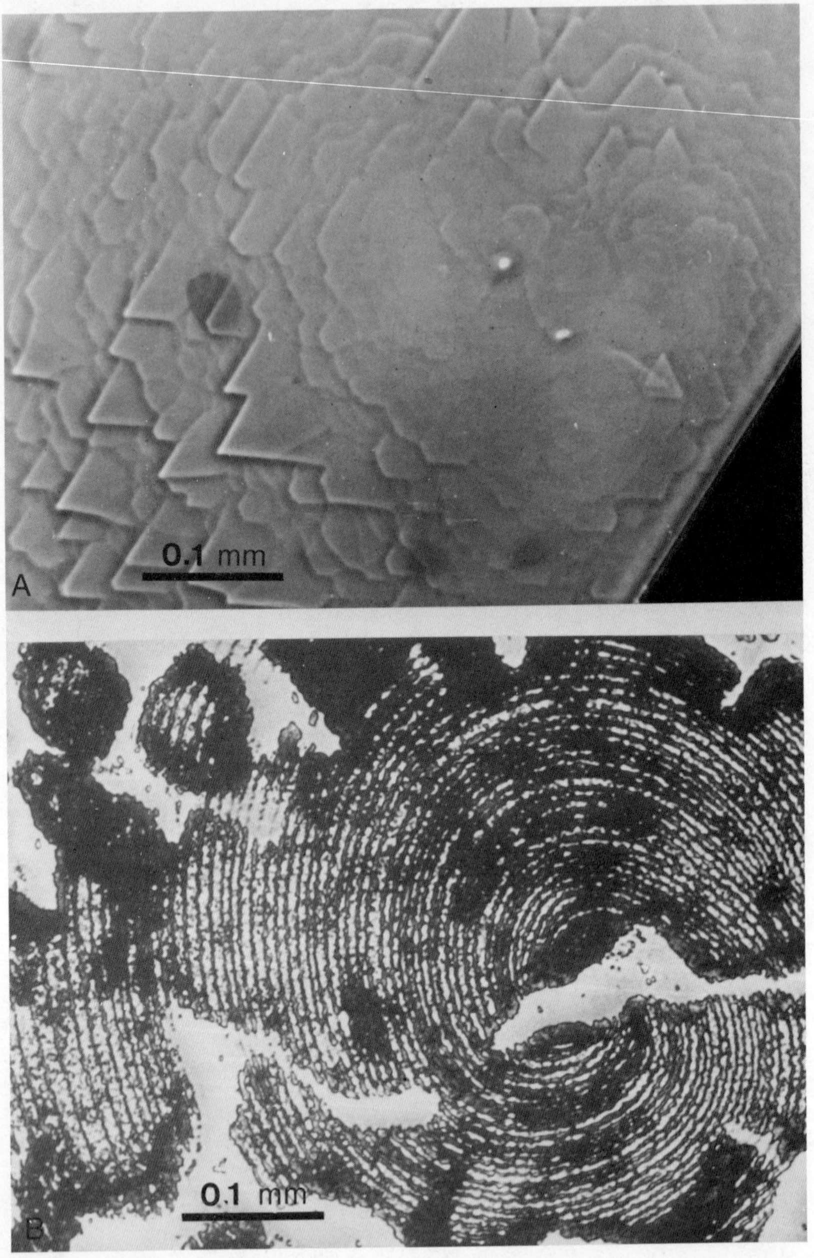

FIG. 36. Showing the difference of the degree of polygonization of spirals on different faces of the same crystal: (A) polygonized spiral on (0001) face; (B) less polygonized spiral on $(10\bar{1}1)$ face of a hematite crystal. Note denticulated spiral steps (see Sec. III.C.5).

show different degrees of polygonization, in addition to the different symmetries described above. Figures 33 and 36 offer such examples. The (0001) face of this hematite crystal exhibits almost polygonized triangular growth spirals, whereas the $(10\bar{1}1)$ face on the same crystal shows almost circular spirals. The $(10\bar{1}1)$ face is morphologically a less important face, and therefore its ξ with respect to that of the (0001) face is less than unity. A similar difference among different crystal faces on the same crystal is also seen in Fig. 33. The morphologically most important (0001) face exhibits well-polygonized spirals, the less important $(10\bar{1}0)$ face has polygonized spirals with rounded corners, and the least important $(21\bar{3}1)$ shows circular forms.

5. *Rough Spiral Steps*

As already discussed in Sec. II.B.6, undulation of the order of about 500 units in length and 30 units in width should be expected for spiral steps for $\alpha = 4$. These values correspond to 1000 and 100 Å, respectively, for steps on the (0001) face of SiC or hematite. However, within the limit of lateral resolution of optical microscopes, steps of ideal growth spirals are in most cases very smooth. This is also true for most growth spirals observed by electron microscopy, as can be seen in Fig. 23.

There is, however, only one exceptional case in which undulation of spiral steps was noticed universally, i.e., on hematite crystals from Sasazawa, Nagano Prefecture, Japan. These crystals are formed by a natural chemical vapor transport process due to postvolcanic action, with Cl as the transport agent. Rather high-temperature conditions as compared to hematite crystals from other localities are estimated for this locality, judging from their geological occurrence and also from the common occurrence of more circular growth spirals than those observed on the crystals from other localities. This suggests lower α values for the crystals of this locality. In this case, an optically detectable undulation is seen.

For ideal monomolecular growth spirals as exemplified in Fig. 37 A, the degree of undulation becomes more exaggerated when a large number of spiral layers originates from a group of dislocations (Fig. 37 C) or as monomolecular growth layers bunch together to form thicker layers. These cases are shown in Fig. 37 C and D. As can be seen

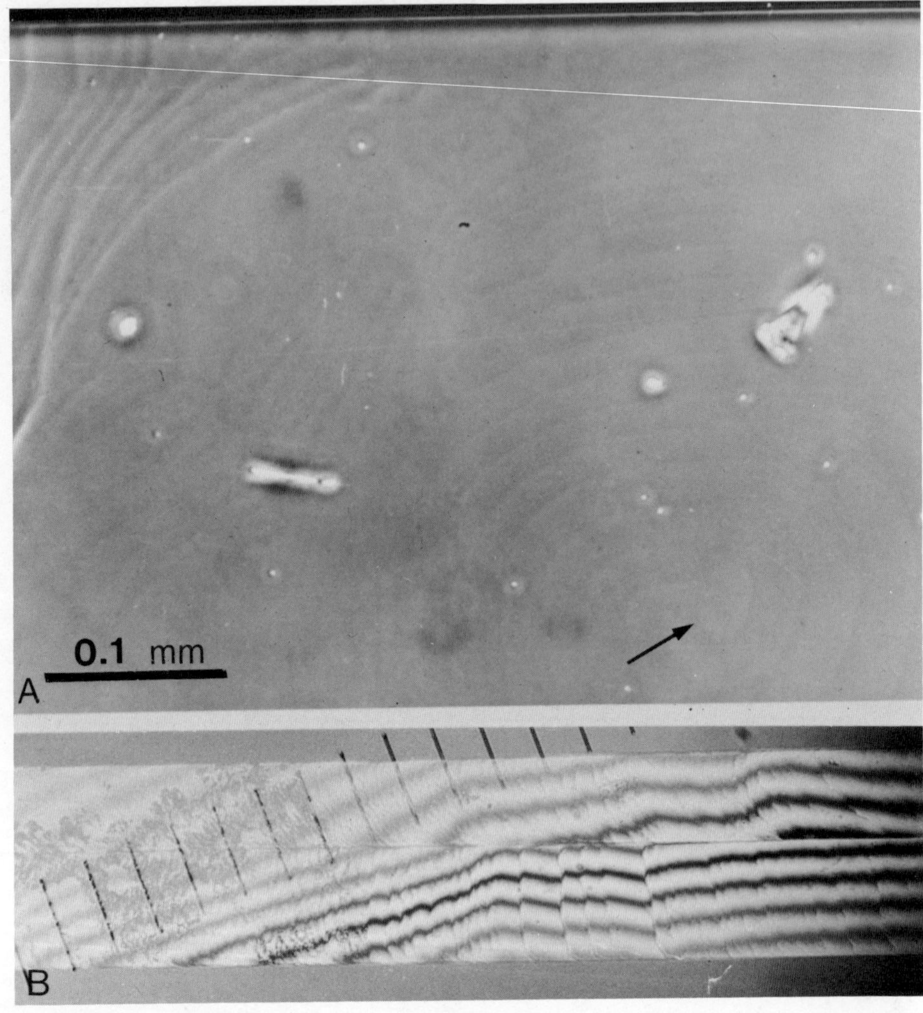

FIG. 37. Showing undulated step of ideal monomolecular growth spiral, whose center is indicated by the arrow (A), and of bunched thick layers (C), (D) observed on the (0001) face of hematite from Sasazawa, Nagano Prefecture, Japan; (B) is a two-beam interferogram to show the mode of bunching and the heights of bunched layers. Also note the increase of wavelength of undulation as step height increases.

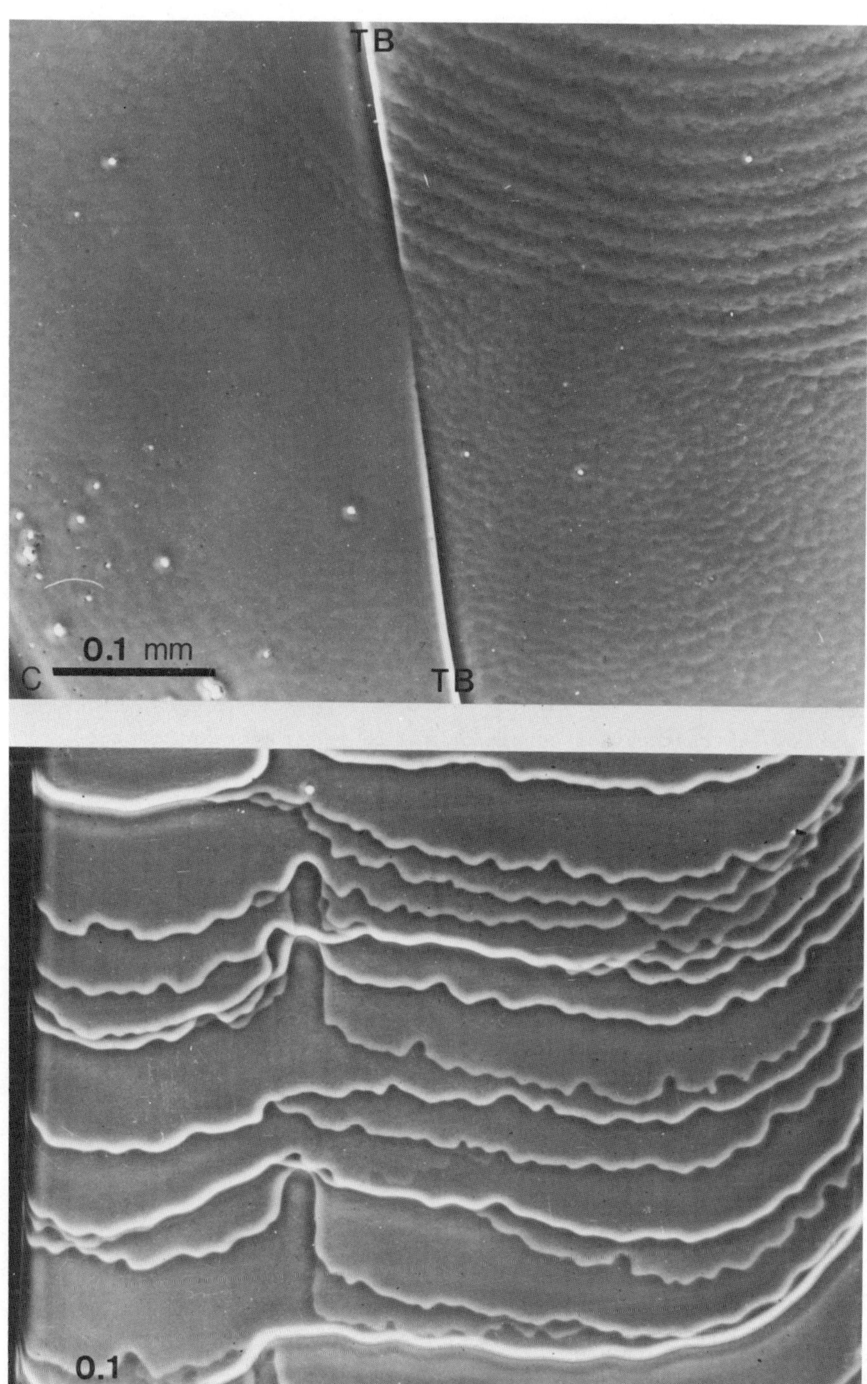
TB
0.1 mm
C
TB
0.1
D

from these photographs, the wavelength of the undulation becomes longer as the steps become thicker and as the chance of contact with successive layers increases. These observations, especially on the change in wavelength of the undulation due to bunching, are in good agreement with the theoretical expectation discussed in Sec. II.B.6.

As discussed in Sec. II.B.6, a rough step may occur either by a compound mechanism of spiral growth and two-dimensional nucleation growth or for three other possible reasons: (1) decrease of λ_0 due to groups of dislocations; (2) special kinetic reasons; and (3) occurrence of surface diffusion. Since the measured λ_0 values on single spirals on these crystals are on the order of 20 μm, which suggests a very low β, it is highly improbable that a two-dimensional nucleation mechanism has taken place simultaneously with spiral growth. A compound mechanism is therefore excluded from the possible origins for the rough steps of these crystals. Since highly complicated factors are involved in natural crystallization, it is quite difficult to claim a single origin for the rough steps. It is suggested, however, that a strong solid-fluid interaction and a high temperature and low supersaturation condition may be very important factors.

D. Deviation from Ideal Spirals

1. *Effect of Stress Field*

In 1951, Frank [75] predicted that if the Burger's vector of a dislocation exceeds a certain critical value, say 10 Å, it creates a hollow core to compensate for the strain energy of the dislocation by creating a free surface. Cabrera and Levine [44] introduced a strain term into the equation of step velocity of a spiral given by BCF [1],

$$v = v_\infty \left(1 - \frac{r^*}{\rho} - KR_0 \frac{r^*}{R^2} \right)$$

where v is the advance velocity of a circular step, v_∞ the advance velocity of a straight step, r^* the radius of two-dimensional critical nucleus, ρ the curvature of the step, R the distance from dislocation to the step, R_0 the radius of hollow core defined by Frank [75], and $K = \gamma'/\gamma$, the ratio of surface to edge free energies.

Recently, three papers have been published on the effect of stress field on the morphology of growth spirals [47,55,56]. In Ref. 55, theoretical considerations were made of the thermodynamic conditions for the occurrence of stable hollow cores by the stress of dislocations. It was demonstrated that eight different cases can be distinguished, which determine whether a stable or metastable core occurs or is absent. This difference is caused by special combinations of a parameter f, determining the shape of the strain functions, and the super- or undersaturation. In Ref. 47, the influence of stress fields on the shape of growth and dissolution spirals was discussed, starting from Cabrera and Levine's formalism. It was shown that (1) although growth and dissolution spirals are symmetrical when there is no strain, both become asymmetrical when strain is introduced; (2) growth spirals show a change of curvature of the spiral steps, which results in the formation of hollow cores when strain exceeds a critical value at the spiral center, whereas dissolution spirals show no change of curvature and form trumpet forms; and (3) the amount of change of curvature varies depending on the type and amount of strains. In the third paper [57], a series of observations on the change of curvature and the formation of hollow cores on growth spirals was presented and interpreted on the basis of the two preceding theoretical papers.

Hollow cores at the center of growth spirals were first observed by Verma [3,76], followed by Amelinckx [4] on SiC crystals, and by Forty [77] on CdI_2 crystals. They offered these hollow cores as examples of those discussed by Frank [75]. They did not notice the change of curvature at the hollow core, though their photographs clearly show this. According to the theories [47,55], the curvature of a spiral step at the center changes from negative to positive. Thus a hollow core is formed at the center of a growth spiral if stress field is convex, i.e., $f > 0$, which causes an accumulation of stress in the origin of the cylindrical stress field. Although in this case a stable hollow core can principally occur irrespective of the strength of the stress field, a hollow core with diameter less than 5 Å is similar to its absence from a physical point of view. The hollow core and the change of curvature at the center become observable under the optical microscope only when the diameter of the core exceeds the resolution of the microscope. If $f \geq 0$, which corresponds to a concave stress field or a constant stress field, a hollow core does not appear at the spiral center but the spiral step changes its curvature from weakly positive to weakly negative, through an almost straight step, depending on the strength of the stress field. Depending on the strength and the state of the stress field, various states may appear at the spiral center. Examples corresponding to all these states have been observed on actual crystals and only representative examples are shown below.

Calculations on the relation between the Burgers' vectors and the radius of equilibrium hollow cores using Frank's formula [75] have shown for SiC that growth spirals with step height of less than about 50 Å show neither observable hollow cores nor change of curvature. In fact, on growth spirals with a step height less than 20 Å, neither hollow core nor change of curvature is detected, as seen in Fig. 24. In the case of spirals with step heights higher than about 80 Å, hollow cores and change of curvature are in general observable on SiC, provided the spiral is not perturbed and has an almost Archimedean form. Figure 38 shows a representative example. The diameter of the hollow cores depends principally on the height of the spiral layers, but can be perturbed considerably due to various factors, such as exchange of spiral centers or the state of stress field. Figure 39 shows an example of the former case, where the diameter depends principally on the step height (the higher the step, the brighter and the wider the bright halo appearing along the step), but at the same time the exchange of spiral centers in group b gives

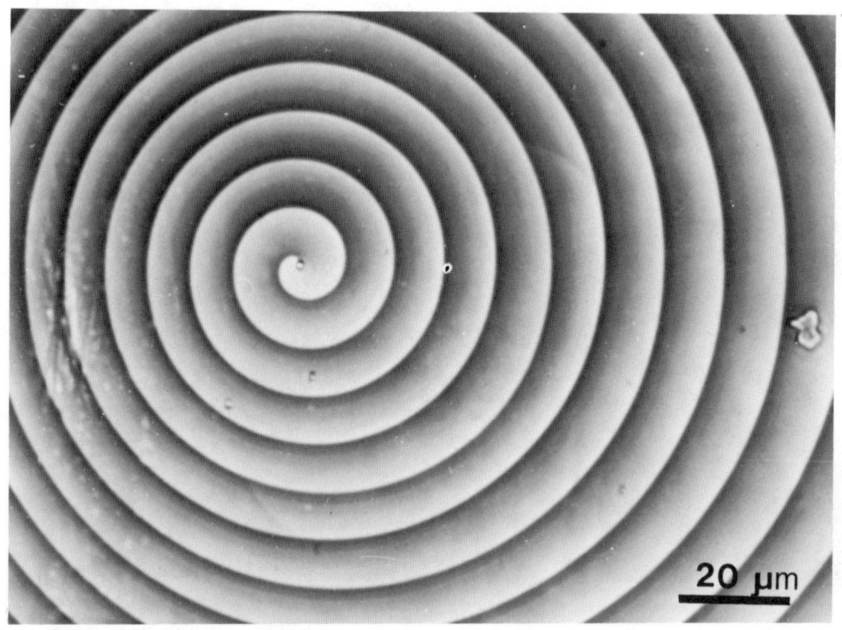

FIG. 38. Growth spiral with hollow core at its center; (0001), SiC. The step height is about 90 Å. Also note the change of curvature of the spiral step from positive to negative at the spiral center.

a special effect on the mode of change of curvature, which is schematically shown in Fig. 40.

When a large number of screw dislocations concentrate in a small area, the nature of the stress field can be modified to a concave type, i.e., $f \geq 0$. In this case spiral steps may show a change of curvature but do not form hollow cores at the centers. Figures 41 and 42 are good examples. In Fig. 41, we can see only a slight change of curvature from positive to negative near the origin but no hollow core except at the spiral indicated by arrow. In Fig. 42 the composite spiral shows a depression at the center and a change of curvature near the center, but no hollow cores are formed.

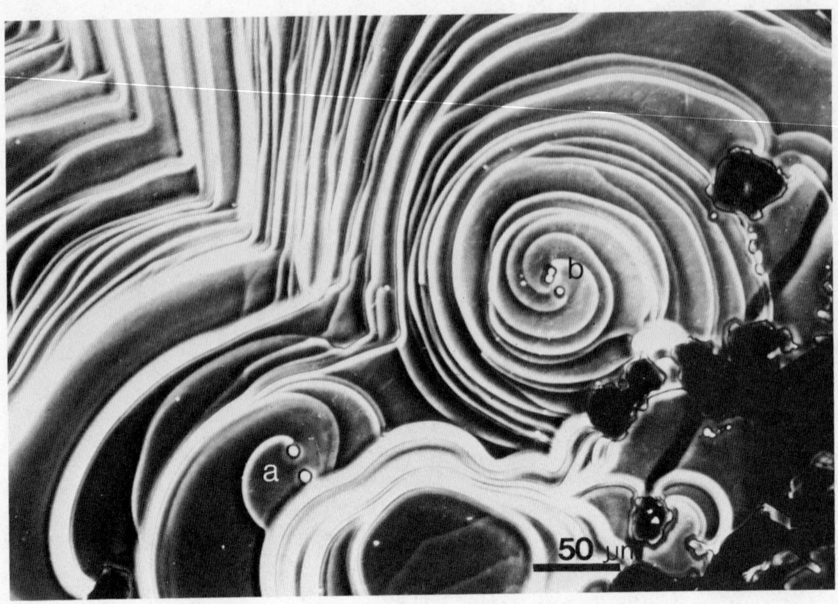

FIG. 39. Showing variation of diameters of hollow cores depending on the step heights; (0001), SiC. Spirals in b group show larger diameters of hollow cores than expected for their step heights, which is due to the exchange of spiral centers during growth.

Stress fields which can influence the advancing of spiral steps need not be associated with dislocations. Any kind of stress field influences the advancement of growth steps. Stress fields associated with a twin boundary, a stacking fault, or misoriented crystals attached to the growing crystal, etc. can cause the retardation of growth layers. Figure 43 shows such an example. The triangular island is formed by epitaxial settlement of a platy crystallite on the growing surface in a twin orientation with respect to the host [78], which results in the introduction of a stacking fault. This was proved by the $2.\bar{3}$ Å level difference between the two surfaces

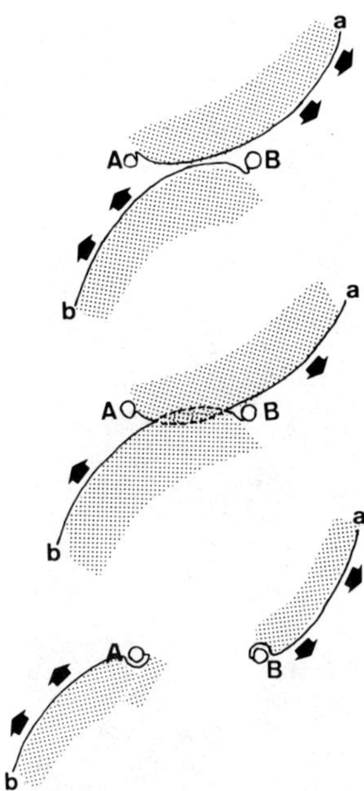

FIG. 40. Schematic drawings to explain the increase of apparent diameters of hollow cores due to exchange of spiral centers. (From Ref. 56.)

[78]. Because of this stacking fault, a stress field forms around the triangular island, which influences the advancement of spiral steps of 7 Å on the surface of the host crystal. The spiral steps coming from the upper part of the photograph are retarded in their advance around the island, forming a pattern similar to a stream around a rock in a river. As soon as the surface of the $\bar{7}$ Å spiral reaches the same level as the island, it can now advance over the surface of the island, resulting in the appearance of tonguelike terrace, as shown in Fig. 44. Thus a stacking fault can cause bunching [79].

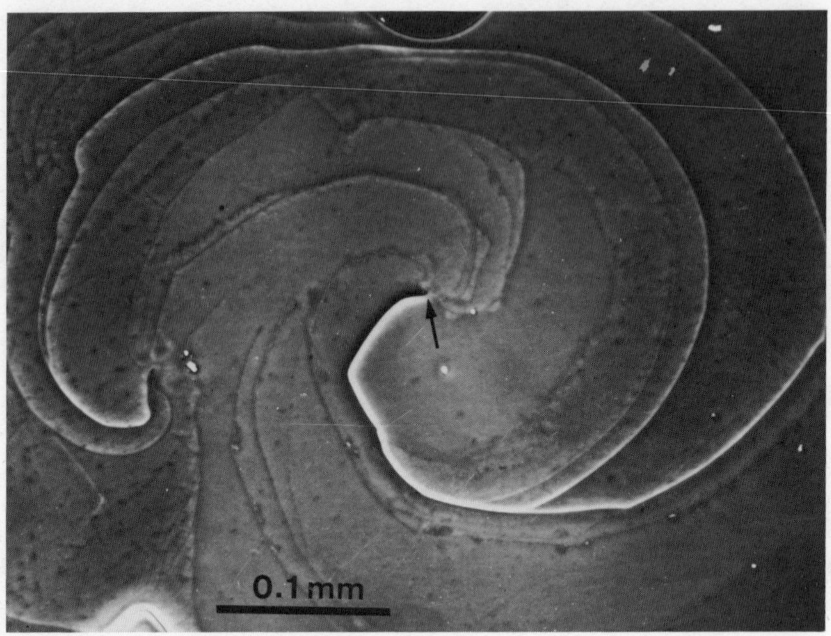

FIG. 41. An example showing a change of curvature from positive to weakly negative without creating a hollow core; (0001), SiC. Only the spiral with the highest step height forms a small hollow core, as indicated by the arrow. Other spirals show only a change of curvature.

Foreign crystals attached to the growing surface have been observed quite universally on the advancement of spiral steps.

When a twin boundary emerges from the common surface of a twinned crystal, thin spiral layers developing on the surface universally show kinks where they cross the twin boundary. This is typically seen on the common (0001) face of twinned crystals of hematite with $(10\bar{1}0)$ twinning plane [80], and can be attributed to the stress field associated with the composition plane of the twin. Figure 45 shows a typical example wherein kinks of spiral steps are seen along the twin boundary.

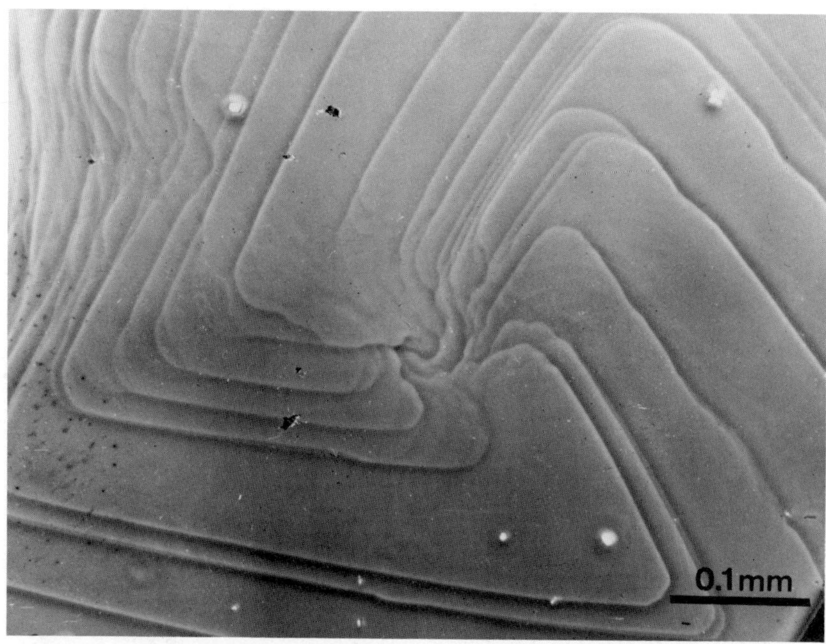

FIG. 42. Formation of a depression at the center of a composite spiral. Accumulation of screw dislocations in a small area creates a concave stress field, but not a convex stress field. As a result, only the curvature of spiral steps changes, but hollow cores are not formed. (0001), hematite.

2. *Effect of Nonuniform Supersaturation Distribution over the Surface*

Since the supersaturation over a growing surface is not uniform due to various factors, the morphology of growth spirals usually deviates considerably from the ideal Archimedean form, in which an uniform supersaturation is assumed. Various factors may cause deviations of the spiral morphology from the Archimedean form. Only a few representative examples are shown in the following.

The most impressive example is the observation of eccentric growth spirals on SiC crystals synthesized by the Lely method [58]. In the Lely method, platy crystals of SiC grow standing from the wall

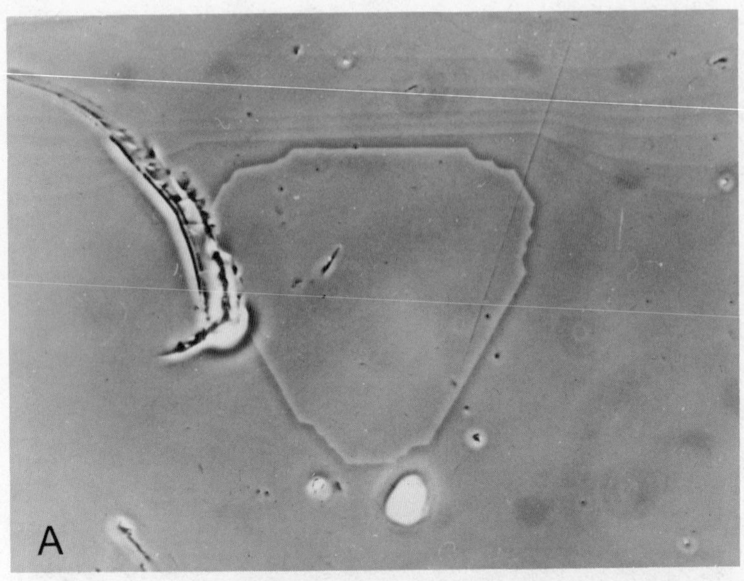

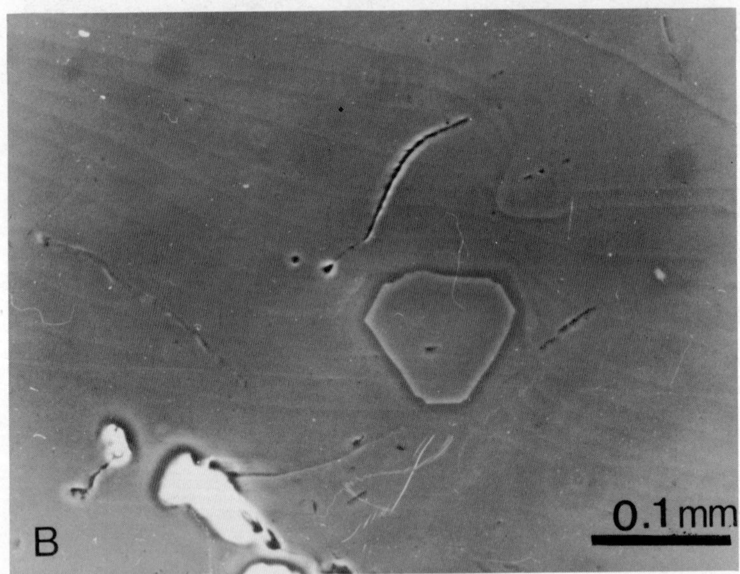

FIG. 43. Modification of advancement of spiral steps of 7 Å height around triangular platy crystallites which settled on the surface of the host crystal in a twin orientation. Showing two stages in (A) and (B). (0001), hematite.

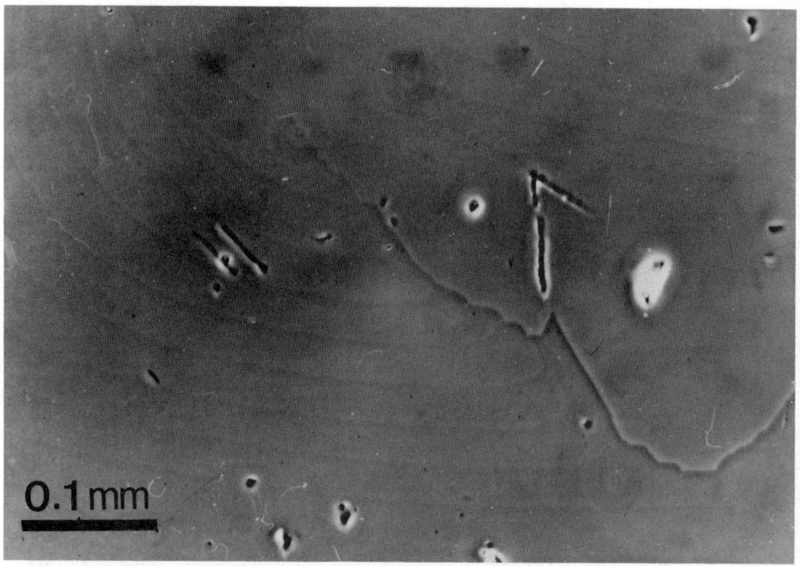

FIG. 44. Tonguelike terrace formed on the same surface as Fig. 43, by further advancement of the spiral steps.

of a crucible made of SiC powder. There is a temperature gradient from the hotter wall to the cooler inner portion of the crucible. A supersaturation gradient is therefore assumed over the surface of a growing crystal, in which the supersaturation is lower at the base of the crystal than the portion far from the wall. Where a higher supersaturation is assumed, the spiral steps advance more rapidly and so a wider step separation is expected, whereas narrower step separation is expected where a lower supersaturation is assumed, provided that the spiral steps originate from the same screw dislocation which is more or less fixed. This results in an eccentric growth spiral, with wider step separation near the edges of the face situated far from the crucible wall, and a much narrower step separation near the base. Such eccentric spirals are in fact commonly observed on the SiC crystals synthesized by the Lely method, with a representative example shown in Fig. 46 [58].

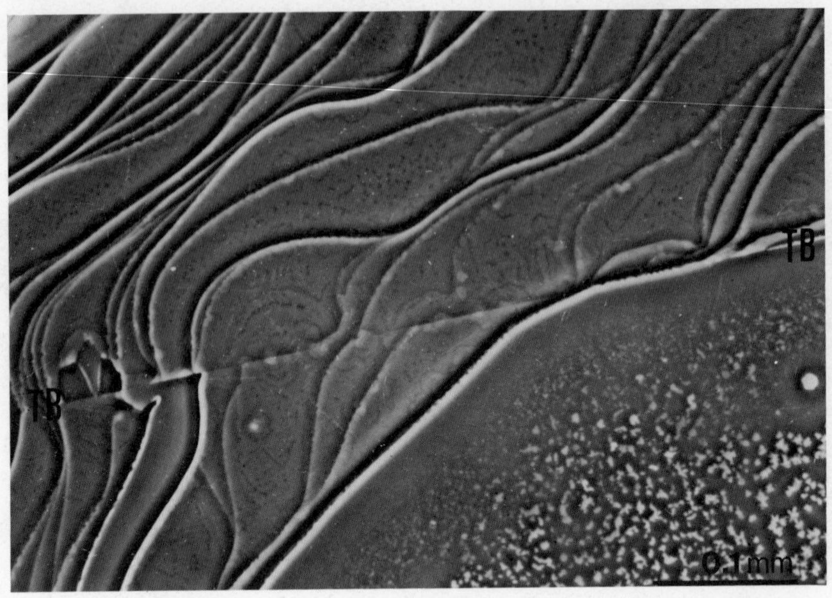

FIG. 45. Thin spiral steps show kinks at the twin boundary (TB); (0001), hematite.

It is well known that supersaturation over a growing surface is not uniform. Edges and corners protrude into a higher supersaturation region, and the center of the face is at the lowest supersaturation. This is often called the Berg effect or Berg phenomenon [81]. Due to this effect, spiral layers advance more rapidly along the edges of a face causing a wider step separation than elsewhere. Such deviation has been observed on many crystals, with an example shown in Fig. 47.

However, it is also sometimes observed that the step separation of a spiral originating from the central portion of a face becomes narrower as it approaches the edges. Figure 48 is a typical example showing such surface microtopographs observed on a kaolinite crystal. This is attributed to a morphological effect, i.e., a much slower growth rate of the side faces as compared to the advancement of the spiral layer, which results in piling up of spiral layers near the edges of the crystal.

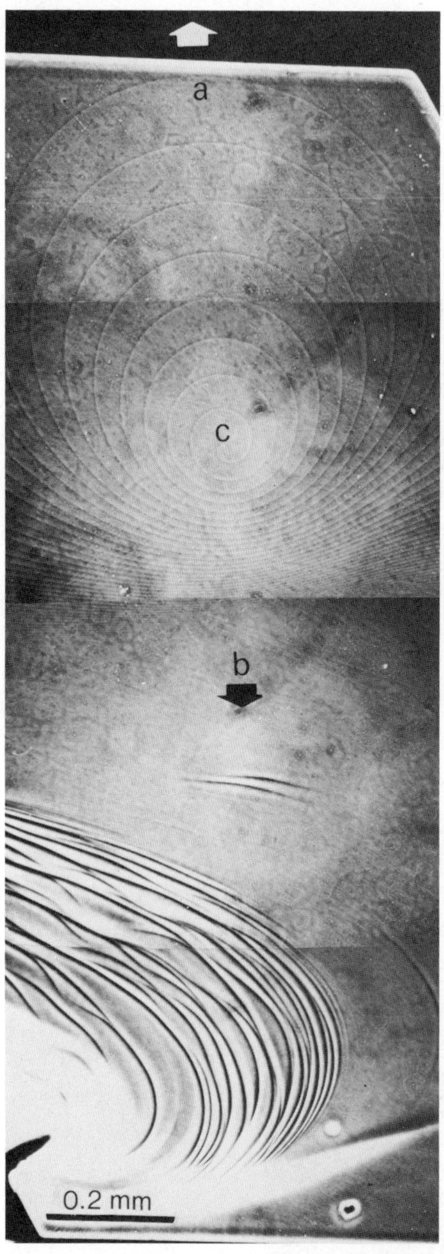

FIG. 46. A representative eccentric growth spiral observed on a SiC crystal synthesized by the Lely method. Step height is 15 Å; (c) indicates the spiral center, (a) the edge of the crystal where a higher supersaturation is assumed than (b). Also note bunching of spiral layers near the base of the crystal. Photo by K. Narita.

FIG. 47. Perturbation of spiral morphology due to Berg effect (higher supersaturation at corner). Note wider step separation along the edges of the crystal. (0001), SiC; interference contrast photomicrograph.

FIG. 48. Electron photomicrograph of a dickite crystal using decoration technique. Note that the step separation diminishes sharply near the edges of the crystal. Also note that the side faces appear by piling up of the steps of spiral layers developing on the (001) face. Photo by Y. Koshino

3. *Effect of Reentrant Corners*

If there is a two-dimensional reentrant corner which appears where two spiral layers meet, this also modifies the morphology of growth spirals, due to the more rapid advance of the spiral steps at the reentrant corner than the steps other than the corner, since a reentrant corner collects more crystallizing entities. Figure 49 demonstrates this effect. It is clearly seen in this photograph that the steps advance much faster at the reentrant corner indicated by arrows than the other directions. Such reentrant corners occur where two spiral steps from different sources meet, or where spiral layers meet at the steps of higher layer or crystal edges. Such a deviation is commonly observed.

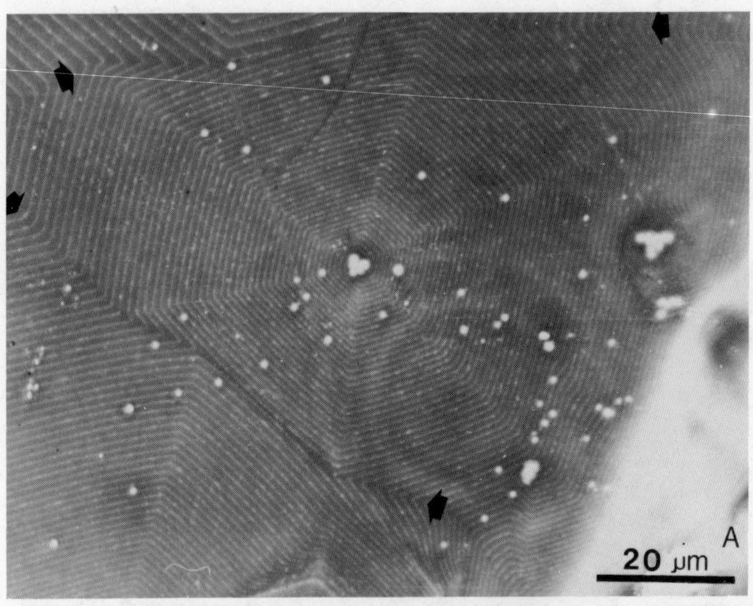

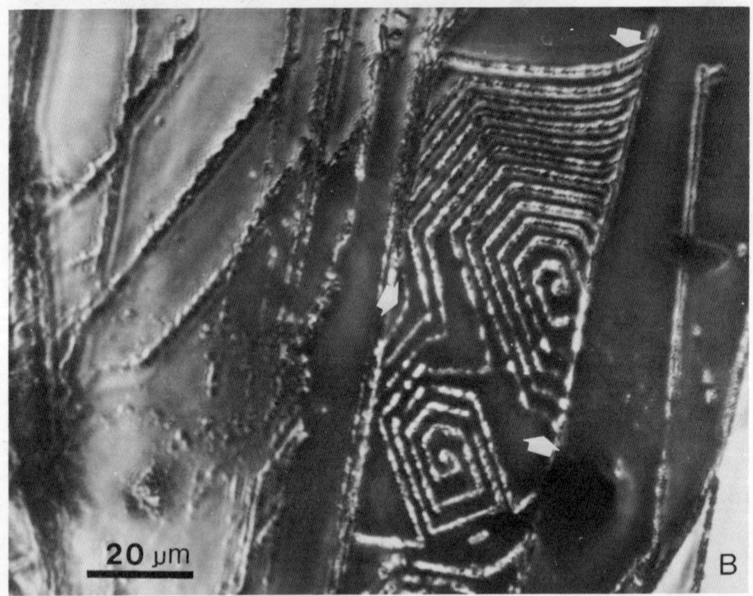

FIG. 49. Effect of two-dimensional reentrant corner on the advancement of spiral layers: the effect observed (A) at the places where spiral steps from different origins meet; (0001), SiC; and (B) between spiral steps and a higher step; (001), phlogopite. Arrows show the directions where the effect is seen.

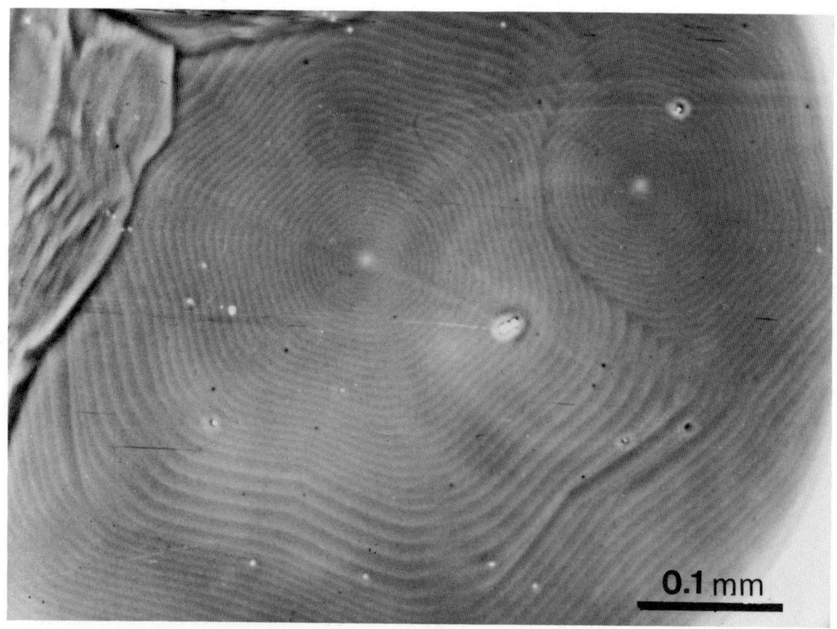

FIG. 50. Starlike spirals, with six corners protruded, due to the more rapid advance of steps in the interlaced directions; SiC, 4H polytype.

4. *Perturbation Due to Interlacing*

Figure 50 shows an almost starlike spiral, with six corners protruded. This crystal belongs to the 4H polytype of SiC, which has a 10 Å unit cell height consisting of two 5 Å units stacked in a zigzag way. As briefly explained in Sec. III.C.4 (see Fig. 35), the spiral layer originating from a single screw dislocation with Burgers' vector of 10 Å splits into two triangular layers rotated by 180°. Due to the difference of the advancing rates of each layer in the same direction, two triangular layers of 5 Å can bunch to form 10 Å steps along the six directions perpendicular to the hexagonal edges, but the two advance independently in the directions of six corners of the hexagon. As a result, an interlacing pattern appears only along the directions of six corners. Since the advancing rate of the thinner steps is higher than that of the thicker steps (see Secs. II.B.7 and III.C.3), six corners will protrude resulting in a starlike morphology. Such deviation is often observed on interlaced spirals.

E. Interaction and Cooperation of Neighboring Spirals

1. *Interaction and Cooperation among the Neighboring Spirals with the Same Burgers' Vector*

Since it was explained in some details in Verma's book [4], we shall describe only briefly the spiral patterns induced by interaction or cooperation between the neighboring spirals. Different types of cooperation or interaction appear depending on the mutual separation, Burger's vectors, the signs of the neighboring screw dislocations, and so forth. The simplest case, i.e., for screw dislocations with the same Burgers' vector, is schematically shown in Fig. 51, which shows the following types of cooperation.

For groups of screw dislocations of the same sign and the same Burgers' vector:

1. When mutual separation of screw dislocations is more than $19r^*/2$, independent spirals appear at first, and after some turns depending on the mutual separation, spiral steps cooperate to form a parallel nonintersecting step system (Fig. 52).

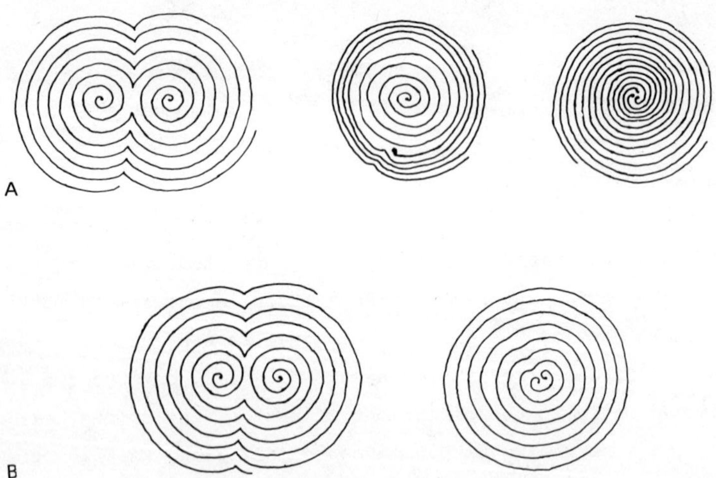

FIG. 51. Schematic drawing of cooperation of spirals originating from neighboring screw dislocations with the same Burgers' vector: (A) with the same sign; (B) with opposite signs.

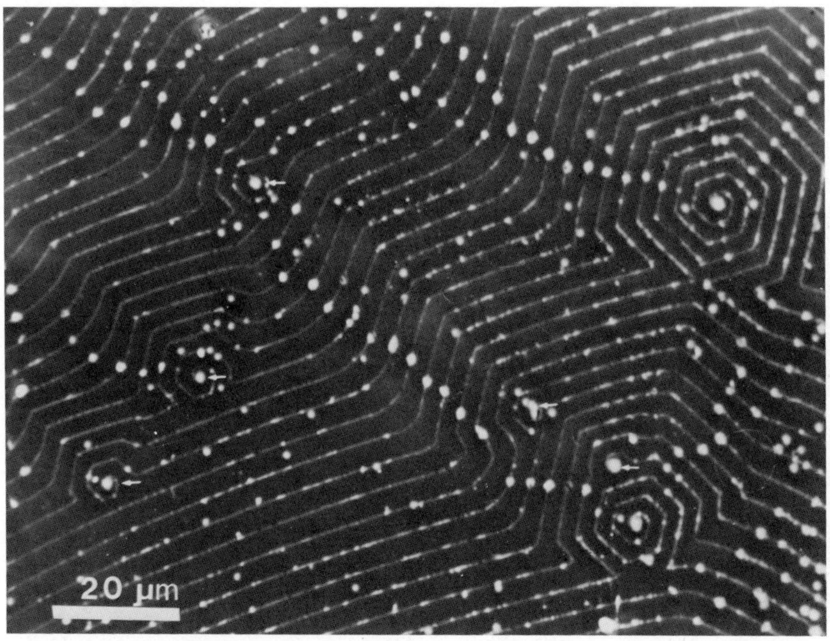

FIG. 52. An example of cooperating spirals originating from screw dislocations with a mutual separation larger than $19r^*/2$; (0001), SiC. Dominated spirals are also seen as indicated by white arrows.

2. If the mutual separation of screw dislocations is less than $19r^*/2$, a parallel nonintersecting spiral step system occurs, which is called a composite spiral. In this case, an exchange of spiral centers is necessary at every turn of the spiral, as discussed in BCF [1] (Fig. 53).
3. Depending on the mutual separation and the time delay in the appearance of spirals, there can be cases in which one spiral shows an ideal form, whereas the other shows no turns at all, but only emits a step parallel to the system of the former, as indicated by arrows in Figs. 52 and 53. The former is called a dominant spiral, the latter a dominated spiral.
4. For a group of screw dislocations of opposite signs and the same Burgers' vector, a closed loop is formed from the beginning

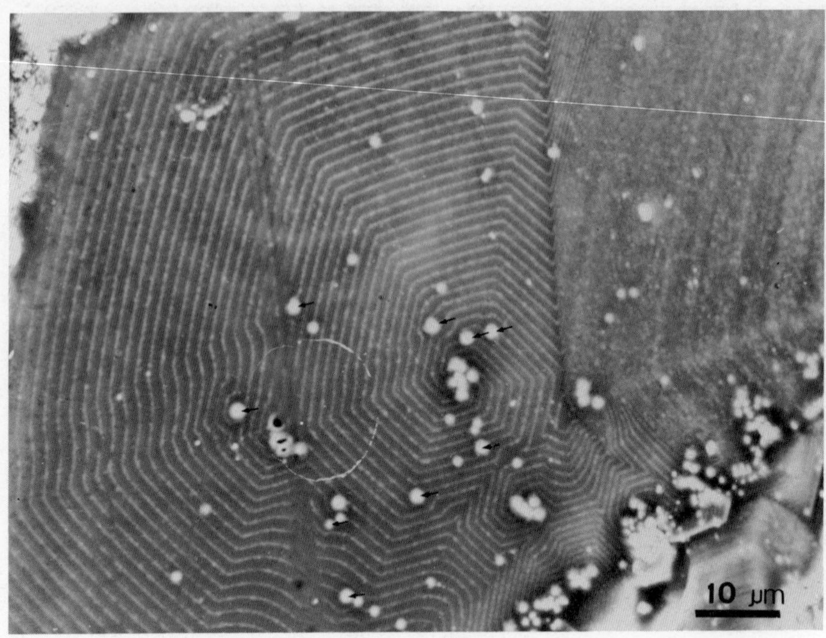

FIG. 53. An example of a parallel step system originating from screw dislocations with a mutual separation smaller than $19r^*/2$. Dominated spirals are shown by arrows. White dots are impurities selectively adsorbed at both screw and edge dislocations intersecting the surface.

or after a few turns depending on the mutual separation (Fig. 54). Again a perturbation can occur depending on the mutual separation and the Burgers' vector.

2. *Interaction between Spirals with Different Step Heights*

If two spirals with different step heights occur near one another, a fault line appears along the boundary between the two territories. If one spiral has a step height twice that of the other, a fault line appears every two layers, and alternating layers form a common surface. If one layer has a step height n times the other, every nth surface forms a common surface, while others show a fault line (Fig. 55).

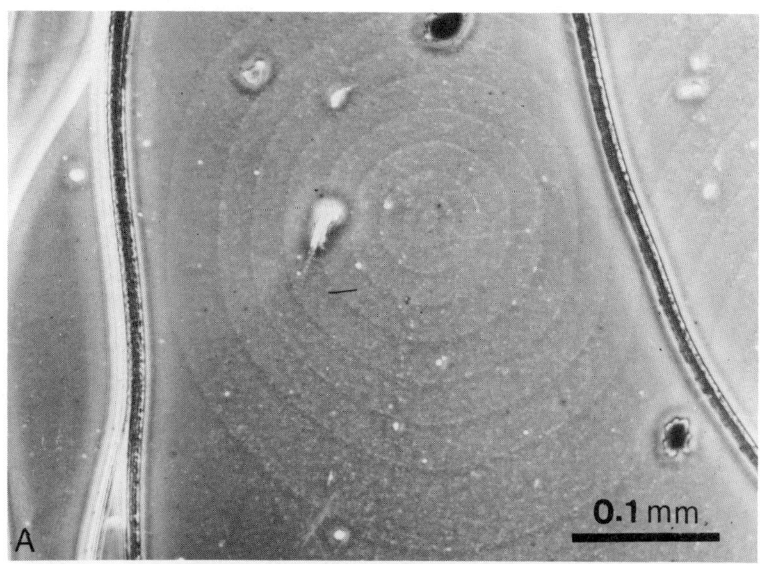

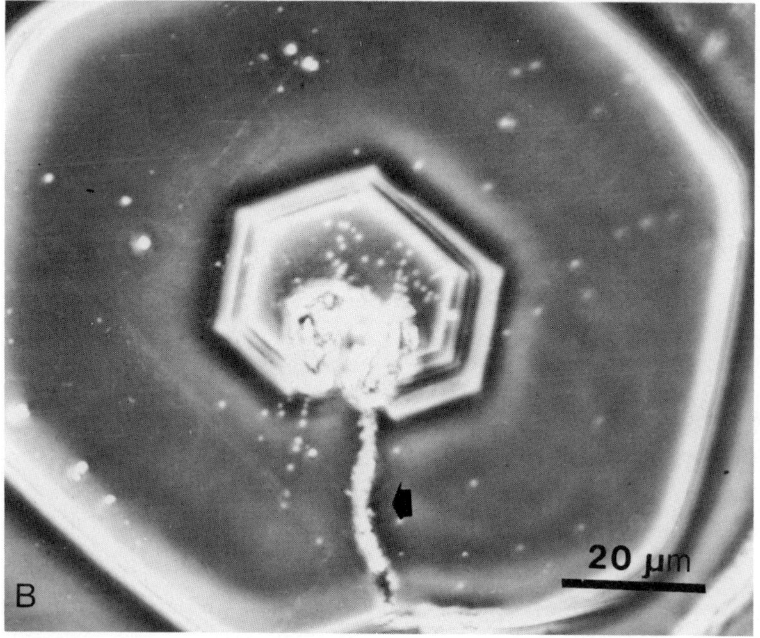

FIG. 54. Two examples of closed loops formed by spiral growth of opposite signs: (A) from two dislocations of the same Burgers' vector; (B) with different total amounts of Burgers' vectors. In (B), a fault line (indicated by an arrow) appears due to the difference in Burgers' vectors between the two groups of screw dislocations.

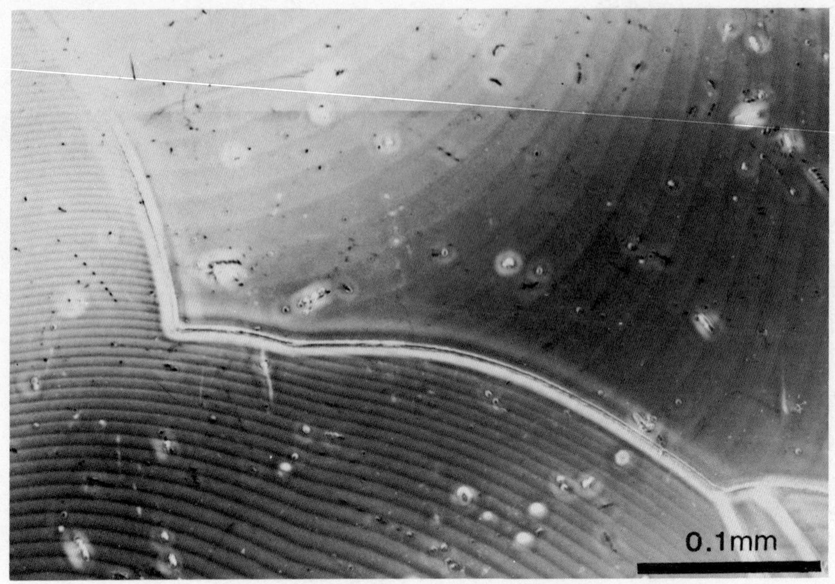

FIG. 55. Fault line appearing along the boundary of two spiral territories; (0001), SiC.

As described in III.E.1, a set of parallel spiral steps with equidistant step separation appears if screw dislocations have equal mutual separation and the crystal grows under a steady-growth condition. In this case, the step separation in the spiral pattern becomes $19r^*/n$, n + 1 being the number of cooperating spirals. If, however, the mutual separation of screw dislocations is not equal, new modes of vibration appear. This is discussed in some detail in Ref. 57, and also in Sec. III.F.2. Figure 56 is a good example of this, in which the composite spirals a, b, and c show equidistant step separation, i.e., with a zeroth mode of vibration. The spiral D shows, in addition to equidistant steps, another set of steps which appear as brighter steps. Such a new mode of vibration is imposed due to the perturbation in mutual separation among the neighboring dislocations. This is called rhythmic bunching. Rhythmic bunching can occur due to other causes [57], about which further discussion will be made in Sec. III.F.2.

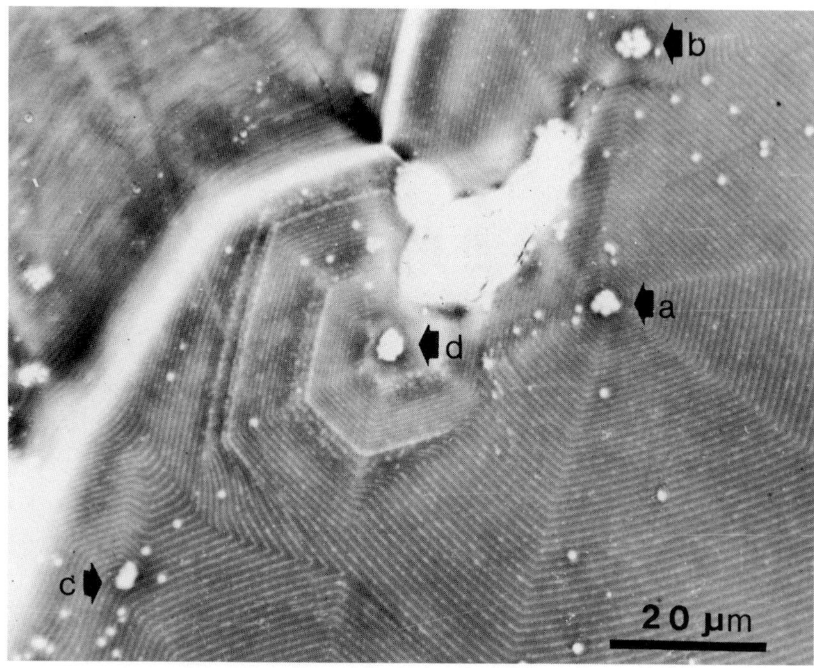

FIG. 56. Parallel step system with equidistant step separation (spirals a, b, and c), and with new mode of vibration (spiral d). White dots are impurities selectively adsorbed at dislocations. Each dot corresponds to a dislocation.

The observations described above all are on the simplest cases, which can be found only on crystals formed under very well-controlled conditions. In reality, both mutual separation and Burger's vectors of screw dislocations are usually not as simple as described above, and the morphology of composite spirals becomes in general very complicated. An example can be seen in Fig. 57. Due to a very small separation between neighboring screw dislocations, the spiral steps bunch to form thicker layers as soon as they emerge, and both thin and thick spiral steps occur together in one composite spiral. Such types are more commonly observed than are ideal Archimedean spirals, especially on natural crystals. It should be stressed here that the main part of a crystal is in general formed by such spirals,

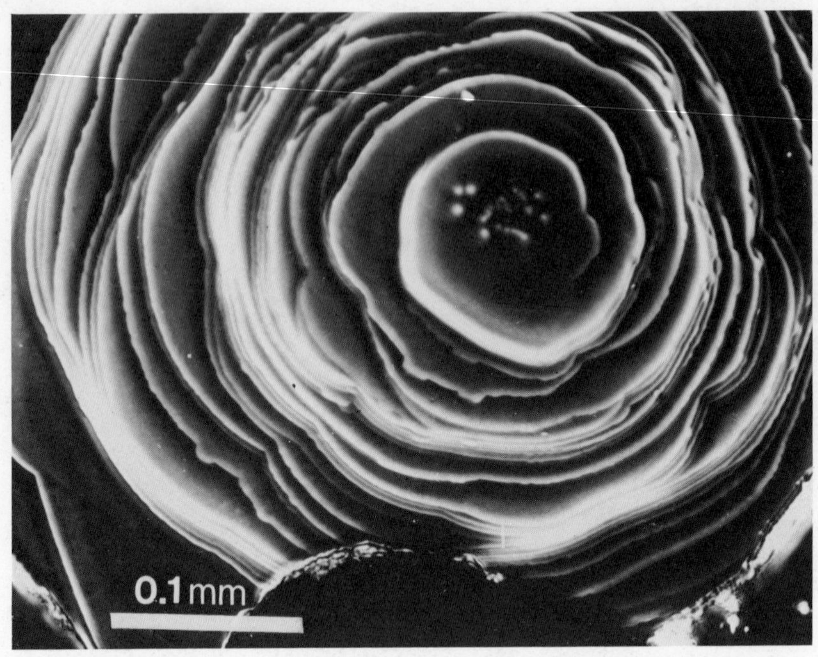

FIG. 57. An example of composite spirals commonly observed on crystal faces; (0001), SiC. At the center of bunched thick spiral layers, a large number of white spots are seen. These are spiral hillocks consisting of monomolecular layers with much narrower step separation. Bunched thick spiral layers seen in the photograph are formed by bunching of these monomolecular spiral layers.

and ideal spirals usually contribute only to the formation of the final topmost surface of a crystal [66].

F. Bunching of Spiral Layers

1. *Elemental Bunching*

Frank [82] proposed a mechanism of bunching of monomolecular spiral layers to account for the occurrence of thick layers of more than 1000 Å. These are thermodynamically unstable if they originate from screw dislocations of large Burgers' vector.

Bunching phenomena are commonly observed during spreading of monomolecular spiral layers. It is worthwhile to describe the initial

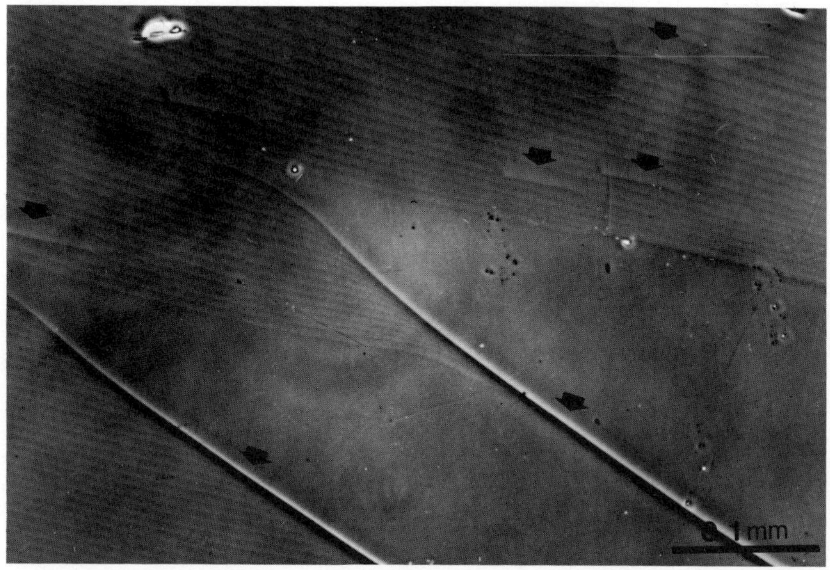

FIG. 58. Initial stage of bunching; (0001), hematite. Single spiral layers have a height of 7 Å and exhibit equidistant step separation. Arrows indicate bunched layers. Also note that the nature of the bright halo changes proportionally to the increase of the step heights. (Refer to the explanation of Fig. 16 and the text.)

stage of bunching during the spreading of monomolecular ideal growth spirals, particularly in relation to the amount of retardation of spiral layers due to the increase of step height, and hence to γ. Figure 58 shows a representative example of bunching, in which thin steps with equidistant step separation are spiral steps of 7 Å originating from a point far outside the photograph. When these steps do not bunch, the step separation is quite uniform. As soon as two steps bunch to form a double step, the advance of the double step diminishes, as can be judged from the step separation indicated by the arrow. In Table 1, the results of measurements of the amount of retardation versus step height on a hematite crystal are shown. This clearly shows that a double step height does not decrease the rate of advancement by a half, but by 0.66 in an average in this particular case. The retardation of rate of advancement due to

TABLE 1
An Example of Retardation of Advance of Bunched Layers, Measured on a Hematite Crystal

Step height, Å	No. of steps	Step separation, μm	Ratio of step separation
7	1	7.5	100
14	2	5.4	72
14	2	6.1	81
28	4	3.6	48
28	4	4.5	60
35	5	4.3	57
273	39	3.9	53
329	47	6.3	85

bunching results in further bunching to form much thicker layers, which are also seen in the photograph.

There can be a variety of causes for bunching. Frank suggested impurity adsorption as a possible cause. Formation of stacking faults, twin lamellae [79], etc. can also be causes of bunching. Figure 59 is an electron photomicrograph of a nacrite (a kind of kaolin group mineral) surface, and shows an example of bunching.

2. *Rhythmic Bunching*

Although possible mechanisms have been suggested theoretically for exciting modes of vibrations in step trains [22,83,84], experimental observations to show how a periodicity is introduced into step patterns have not been reported so far. However, there are cases in our observations to show almost strict periodicity in step trains imposed by modes of vibration. We call such an almost strict periodicity rhythmic bunching [57].

So far three mechanisms have been noticed to cause rhythmic bunching. One type is imposed by the center of the spiral system due to the nonequidistant mutual separation of screw dislocations, which was already discussed in Sec. III.E.2 (see also Fig. 56). A second type is imposed by the systematic decomposition of a regular system of double steps into two single steps, or of a regular system of spiral steps of very large heights consisting of many unit layers

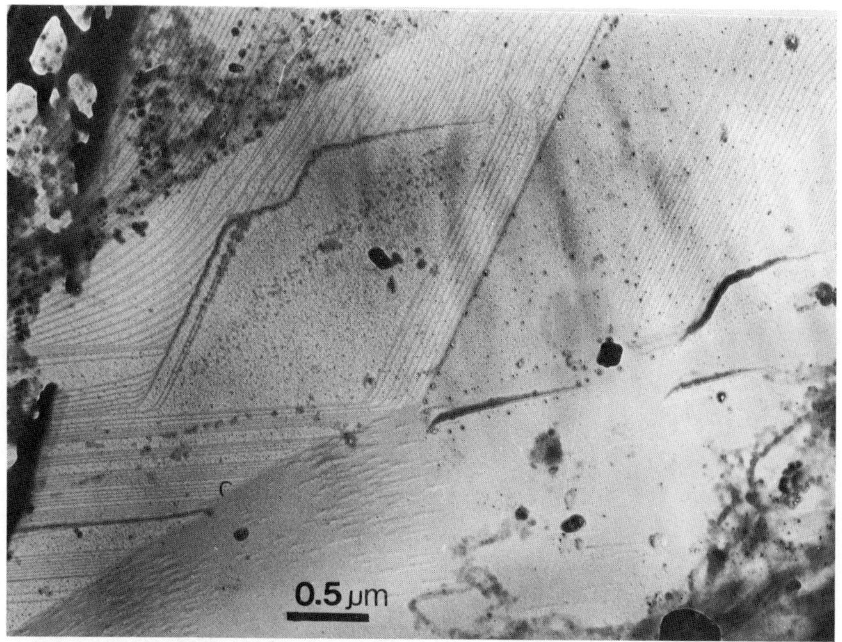

FIG. 59. Electron photomicrograph (decoration technique) of nacrite (a kaolinite group mineral) surface to show bunching due to a stacking fault. Photo by Y. Koshino.

stacked in a zigzag way. Both correspond to the interlaced patterns described in Secs. III.C.4 and III.D.4. Interlaced steps can advance some distance unperturbed.

A third type is imposed by the difference in advancing rate of steps of different heights. Figure 60 shows an example of rhythmic bunching of the third type. The crystal is a 4H polytype of SiC. The spiral originating from a single dislocation with Burger's vector of 10 Å (c_o of 4H polytype) splits into two oppositely oriented triangular layers of 5 Å, forming an interlacing pattern. Due to the quicker advancement of the 5 Å layers, the 10 Å step further decomposes into a pair of $\bar{5}$ Å steps, which are very close together. This makes the $\overline{10}$ Å step in the center more concave with cusps in the center, as indicated by (a) in Fig. 60. After a certain number of turns, a catastrophy occurs due to the fact that the $\bar{5}$ Å layers

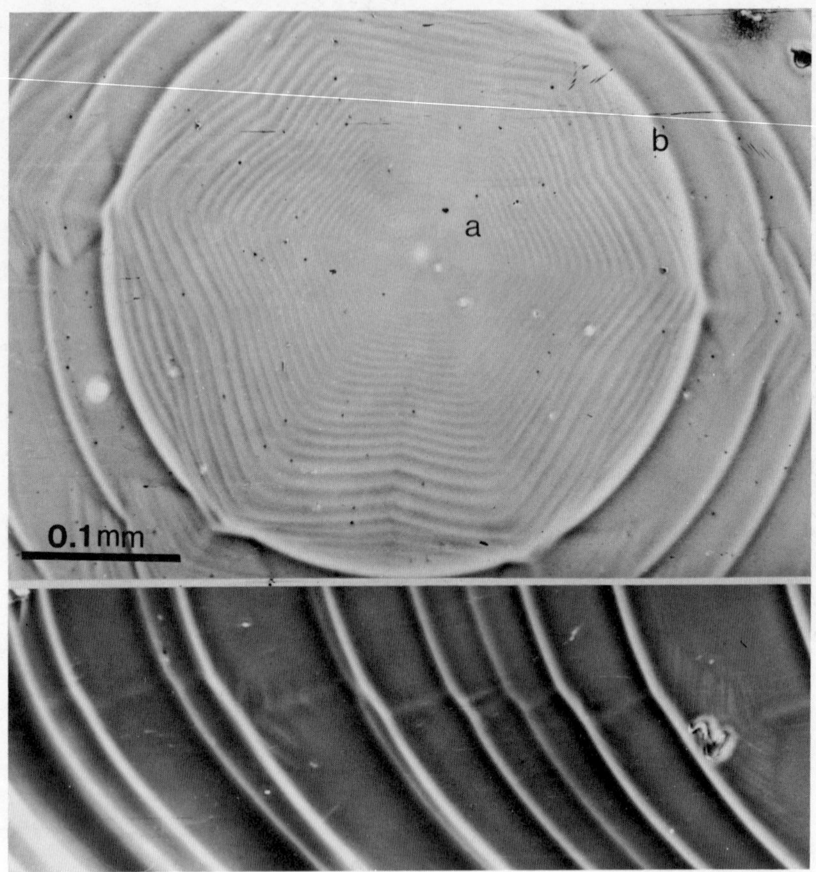

FIG. 60. An example of rhythmic bunching due to the difference in rate of advancement for steps of different heights; (0001), 4H polytype SiC. Circular brighter steps are bunched thick layers, which appear rhythmically as the spiral layers spread outward. The lower photograph shows rhythmically bunched steps in the area away from the center.

bunch together and are delayed. Then the $\overline{10}$ Å layer advances with a higher velocity due to the reentrant corner effect (see Sec. III.D.3), resulting in convex steps, which is shown by b. As soon as both 5- and $\overline{10}$ Å steps align on the same front the same sequence is repeated so that new modes of vibration are created, as can be seen in the photograph.

3. *Realistic Bunching*

However, such an ideal situation is in reality exceptional, and most realistic spirals assume a much more perturbed morphology, due to the bunching of spiral steps. Figure 16 is a good, but still simple, example. In this photograph are seen both thin growth layers originating from a single dislocation and thick spiral layers originating from closely spaced screw dislocations which cannot be resolved under the optical microscope. The white dots are foreign impurity crystallites preferentially adsorbed at dislocation points. Each dot corresponds to one dislocation. It is clearly seen that the thin spiral layer originates from one independent dot, whereas the thick spiral layers originate from a cluster of white dots. A large number of dislocations are so closely together that their spiral layers bunch together to form a thick spiral layer from the beginning of their appearance. Although the thick spiral layers have the appearance of thick layers originating from one independent dislocation with a large Burger's vector, this is not so. Figure 61, which is a more general case, demonstrates the universal occurrence of such bunching while spiral layers are advancing. This type of composite spiral has been observed more commonly on natural crystals than on synthetic crystals.

4. *Effect of Foreign Crystals and Defects*

Since impurities may play an important role in bunching, as Frank suggested [82], we now briefly describe how impurities affect the advancement of spiral layers. Since bunching can occur if there is a factor of any kind which diminishes the advance rate of spiral steps, we also include possible factors other than impurities in the following description.

Decoration of spiral steps by impurities after the cessation of growth has been observed fairly often on both natural and synthetic crystals. Figures 14, 24, etc. show such cases. Decoration increases the visibility of the step considerably, and so steps of about 10 Å become detectable even with an ordinary optical microscope with incident lighting [3,86]. The decoration technique of electron microscopy

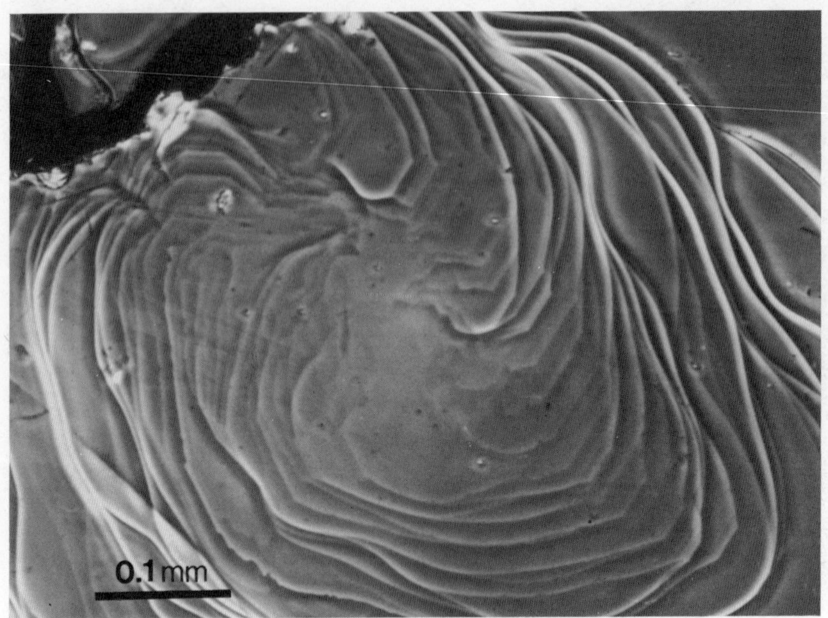

FIG. 61. An example of commonly observed bunched spiral layers in composite spirals.

(see Sec. III.B) utilizes this effect. Impurities also often preferentially adsorb at the points of emergence of both screw and edge dislocations. The white dots seen in Figs. 16, 28, 31A, 49A, 52, 53, 54B, 56, 66, etc. represent such impurities. Both screw and edge dislocations are decorated similarly, and are more pronouncedly decorated than the spiral steps.

We are able to observe the effects of impurities or outcrops of defects upon the advancement of growth layers only when their sizes are larger than the lateral resolution of the microscope used. We are unable to investigate the effect of impurities in the form of ionic entities. Several different kinds of foreign particles or defects have been observed to affect the advancement of spiral layers, and each behaves differently. These include crystals of impurities and crystallites of the same species which are formed in the system.

These may settle on the growing surface in an epitaxial manner, in a twin orientation, or in an inclined orientation. Holes outcropped on the surface, twin boundaries, stacking faults, etc. also influence layer advancement. The effect of foreign crystals, both of impurities and of the same species, varies depending on the degree of similarity between the host and the foreign crystals, i.e., depending on the amount of strain caused by the attachment. When a crystallite of the same species adheres in a crystallographic orientation on the growing surface, growth layers on the host crystal are attracted to the former crystallite, and show pronounced advancement around the latter. When impurity crystals or foreign crystals of the same species adhere on the growing surface in a twin or in an inclined orientation, they play the role of an obstacle to the advancement of the growth layers. Growth layers are repelled around these crystals and thus show concave curvature surrounding them. An example has already been shown in Fig. 43.

A similar type of effect is also seen along twin boundaries on the common surface of a twinned crystal. Growth layers either form kinks where they cross the twin boundary or they cannot cross the boundary, depending on the step heights and the inclination of the step to the boundary. This effect was already discussed in Sec. III.D.1.

Figure 62 shows an outcrop of a hollow tube in a SiC crystal on a growing surface. Advancement of the spiral layers seen in the photograph is not affected by the existence of this outcrop, except where the spiral steps form reentrant corners with the periphery of the outcrop. Judging from the difference in the number of spiral steps counted from a to b on both sides of the hollow, it is safe to assume that the hollow tube has a screw dislocation component with a Burgers' vector equal to 10 layers ($\sim$150 Å). It is interesting that in spite of this, the outcrop of the hollow tube has no effect on the advancement of the spiral steps.

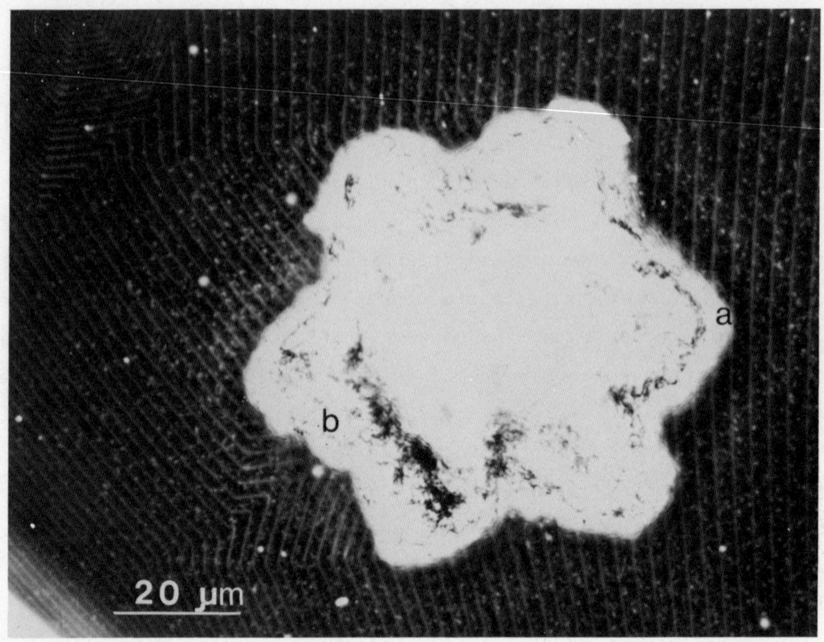

FIG. 62. Mode of advancement of spiral layers around a hollow outcrop on the surface. The hollow has a screw dislocation component with a Burgers' vector equal to a height of 10 layers, which is 150 Å.

G. Actual Surface Microtopographs of Crystal Faces

Since the probability of occlusion of impurities, foreign crystals, or mother liquid into growing crystals increases as crystals grow larger, it is to be expected that the number of spirals developing on a face increases as crystals grow larger. While a crystal is small, the surface may be covered by a spiral pattern originating from a single dislocation. As it grows larger, more spirals appear on the surface. This of course depends on how well the growth conditions are controlled. If crystals grow under very well controlled conditions, a surface as wide as 10 mm^2 may be covered by a spiral pattern originating from one screw dislocation (e.g., Fig. 25). SiC crystals synthesized by the Lely method represent such cases (Fig. 63A). If growth conditions are not well-controlled, the area which

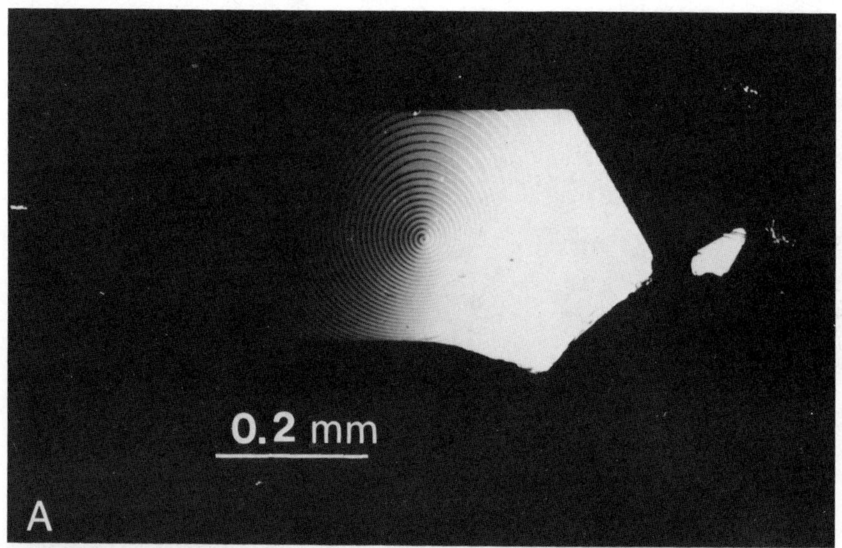
0.2 mm
A

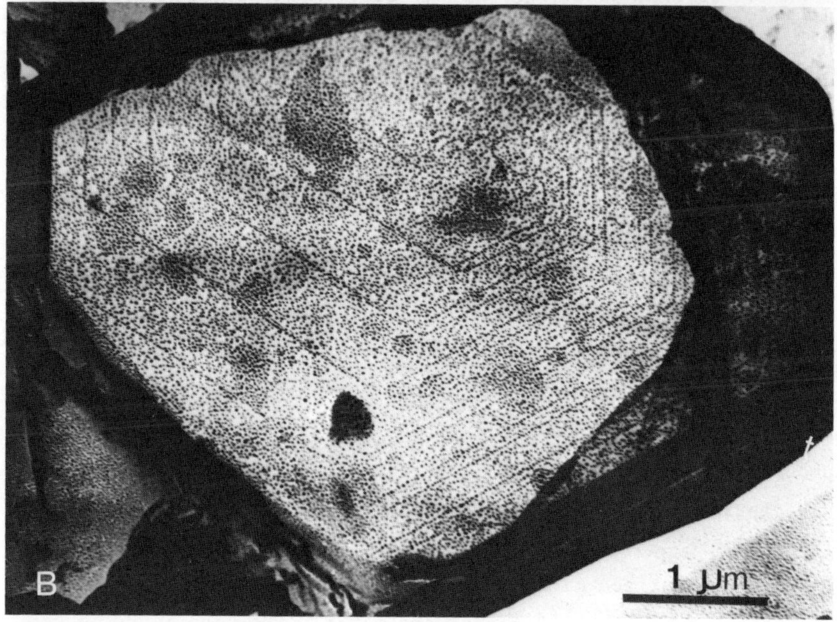
B
1 μm

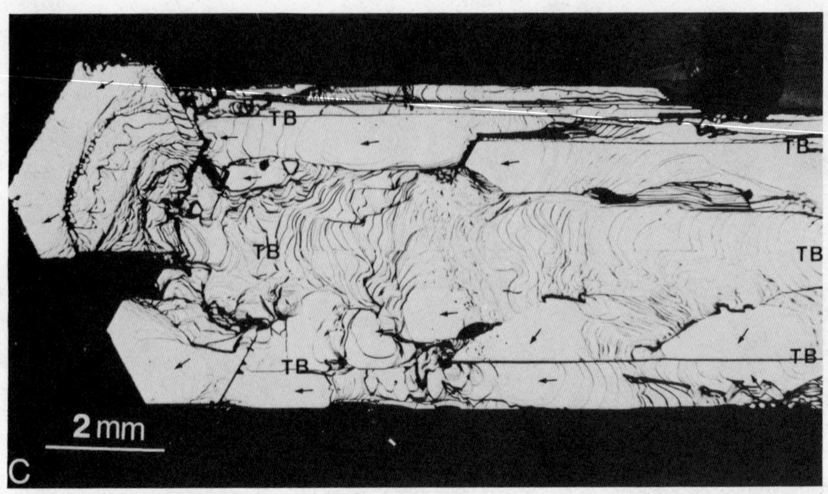

FIG. 63. Three representative examples of actual surface microtopographs of crystal faces; (A) (0001) SiC; (b) (001) dickite; (C) (0001) hematite. Arrows indicate growth centers, TB is a twin boundary. Photos by S. Itabashi and Y. Koshino.

is covered by a single spiral pattern may be as small as 1-10 μm^2 (Fig. 63 B). Many natural crystals grown from the vapor exhibit such behavior. In addition to this, cooperation and interaction of spiral steps from different origins, bunching of spiral steps, and so forth give further complexity to the surface microtopographs of crystal faces (Fig. 63 C). Three representative examples are shown in Fig. 63. The actual surface microtopographs can therefore be very complicated. Yet it is interesting to note that crystals of the same species but from different localities or synthesized at different laboratories or by different techniques do indeed show different characteristics of surface microtopographs which can be used as diagnostic tools to differentiate their different origins [67,70, 85-88]. The surface microtopographs of crystal faces reflect the difference in growth conditions more vividly than one might imagine. To demonstrate this, (0001) faces of natural (in pegmatite), hydrothermally synthesized, and flux-grown emerald crystals are compared in Fig. 64 A, B, and C. The natural crystal exhibits deformed

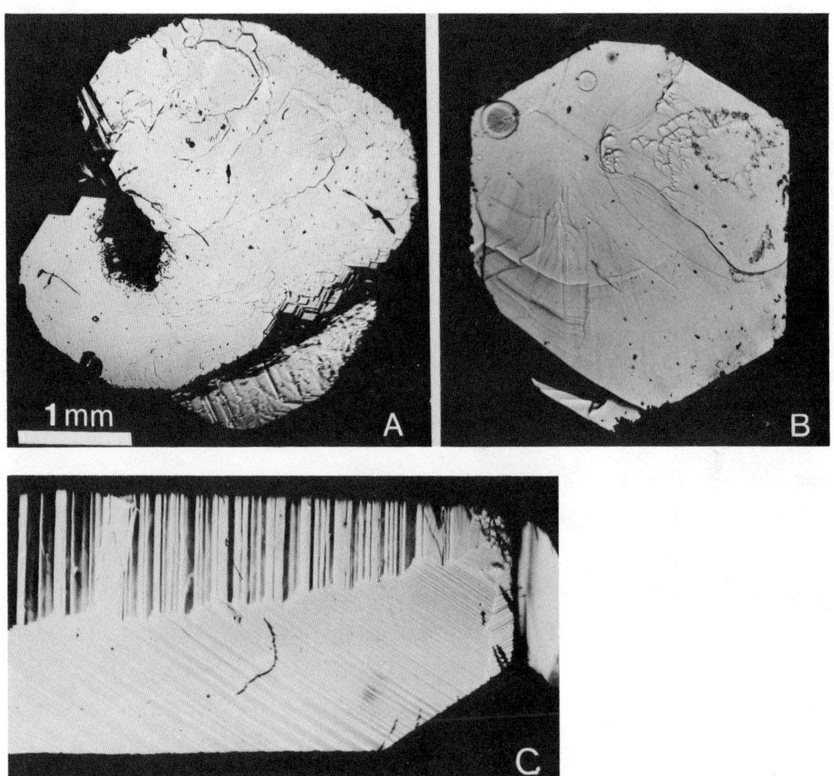

FIG. 64. Surface microtopographs of (A) natural, (B) flux-grown, and (C) hydrothermally sunthesized emerald crystals. (0001).

hexagonal growth spirals with much wider step separation. The hydrothermally synthesized crystal has regular hexagonal growth spirals with much narrower step separation. The flux-grown synthetic crystal shows a large number of circular growth pyramids which consist of very narrowly spaced steps. The step separation is very wide on natural crystals as compared with synthetic crystals. In addition, a clear difference is seen in the number of active growth centers among natural and synthetic crystals, as well as between hydrothermal and flux-grown crystals.

Similar marked differences between natural and synthetic crystals of the same species have been observed on hematite, diamond, garnet, quartz, corundum, etc., and also among crystals of SiC, corundum, phlogopite, hematite, etc. synthesized by different techniques.

H. Comparison of Surface Microtopographs of Crystals of Different Origins

1. *Specimens for Comparison*

In what follows, we make comparisons of the surface microtopographs of crystals of different origins, such as natural and synthetic crystals, crystals grown from different phases, and crystals grown under different conditions.

In comparing surface microtopographs, we mainly focus our attention on the morphology of growth spirals whose step heights are measured, so that their characteristics can be compared with the recent developments in computer simulation of spiral morphology. Although actual surface microtopographs of crystal faces are very complicated, such comparisons are possible and useful because actual surface microtopographs are formed by cooperation and interaction of such spirals.

Of the numerous observations on a wide variety of crystals, the crystals listed in Table 2 are selected for comparison because: (1) enough surface microtopographic observations have been performed to give a clear picture of their general characteristics; and (2) observations on crystals of the same or closely related species grown from different phases and/or under different growth conditions are available.

Among these crystals, natural hematite (b1), phlogopite (d1), and beryl (e1) may be regarded as grown from the vapor phase in natural crystallizations.

In all natural crystallization from the vapor, the vapor is very complex in composition. Vapor growth in nature is therefore mostly by complex chemical vapor transport processes (and not from a pure vapor phase). On the other hand, muscovite (d3) was formed from hydrothermal solutions which reacted with solid rocks and dissolved them.

Table 2

Description of Crystals Observed for Comparison

Crystal	Comments	Growth conditions	Ref.
a. Silicon carbide SiC			
a1. CVT (artificial)	Acheson method, several factories	2000°C	[67]
a2. PV (artificial)	Lely method, Philips and Toshiba	1900-2300°C	[58, 65,67]
b. Hematite Fe_2O_3			
b1. CVT (natural, Cl)	Occuring around volcanic fumaroles from many localities all over the world	400-900°C	[7, 66,78, 80,85, 86]
b2. CVT (artificial)	Synthesized by H. Scholz, temperature-alternating technique; Cl transport	550°C	[92]
b3. HTS (artificial)	Synthesized by J. Remeika, Bell Lab; flux method		[93]
c. Corundum Al_2O_3			
c1. CVT (artificial)	Synthesized by A. Verneuil before his invention of flame fusion method; by sublimation of flux; F transport		[94]
c2. HTS (artificial)	Cryolite flux; by T. Watanabe	1050°C 0.3-2.5%	[95]
d. Mica minerals $X_2Y_{4-6}Z_8O_{20}(OH,F)_4$			
d1. CVT (natural)	Phlogopite 1M polytype occurring in druses of trachyte lavas	<900°C	[96]
d2. PV (artificial)	Fluor-phlogopite, auto-epitaxially grown on cleavage surface of melt-grown fluor-phlogopite; I. Tate	<1410°C	[97]
d3. HS (natural)	Sericite in hydrothermal metasomatic deposits	<200°C	[98]
d4. HS (artificial)	Hydrothermally synthesized muscovite	600°C 1000 atm	[99]
e. Beryl $Be_3Al_2Si_6O_{18}$			
e1. CVT (natural)	Pegmatite		[87]
e2. HTS (artificial)	Flux method; Chatham		[87]
e3. Hydrothermal synthesis; Linde Co.			[88]

Key: CVT = growth by chemical vapor transport; PV = growth from pure vapor phase; HTS = growth from high-temperature solution (flux); HS = growth from hydrothermal solution.

As to synthetic crystals, we may group them into four categories: crystals grown from pure vapor phase (PV) (a2, d2), crystals grown by chemical vapor transport (CVT) (a1, b2, c1), crystals grown from high-temperature solutions (HTS) (b3, c2, e2), and crystals grown from hydrothermal solutions (HS) (d4, e3). Growth temperatures in HS are in general lower than those in PV, CVT, and HTS. Stronger interactions between solid and fluid are expected for HTS and HS than for PV and CVT. Between CVT and PV, the former involves stronger solid-fluid interactions. When natural and synthetic crystals are compared, the natural crystals must have grown from less pure environments and much lower supersaturations than synthetic crystals.

Since we showed in the above individual examples a dependence of morphology of growth spirals on the α factor, on the structure of crystals, and on γ and $\Delta\mu$ through λ_0, we now try to find the similarity and differences among crystals of different origins using the observed morphology of growth spirals in relation to these factors.

Since spiral morphology is controlled by cooperative factors in actual growth processes, it is not easy to correlate the morphology of growth spirals with well-defined growth parameters. However, if a crude simplification is allowed, there are many cases in which the actual morphology of growth spirals is in good accordance with theoretical expectations.

2. *The Results of Observations*

The results of observations are summarized in Table 3, from which several general tendencies emerge as to the difference of surface microtopographs among the crystals grown from different phases and growth conditions. These are summarized below.

1. Between vapor and solution: Crystals grown from vapor phase, both natural and synthetic, as well as both from PV and CVT, invariably exhibit one or two orders of magnitude larger S values (normalized step separation, λ_0/h) than those grown from solution phases, both HTS and HS (e.g., hematite, corundum, mica minerals, beryl). Vapor-grown crystals have normalized step separations of the

TABLE 3

Summary of Observations on the Form of Spirals Found in Crystals Grown by Various Methods

						Mica minerals		
			SiC	Fe_2O_3	Al_2O_3	Phlogopite	Muscovite	Beryl
Vapor	PV		10^4 circular			10^3 circular		
	CVT	natural		10^4 polygonal		10^5 polygonal		10^4 polygonal
	CVT	synthetic	10^3 polygonal	10^3 polygonal to circular	10^4 circular			
Solution	HTS			10^2 polygonal to circular	10^2 polygonal			10^2 circular
	HS	natural					10^2 polygonal	
	HS	synthetic					10^2 circular	10^2-10^3 polygonal

Key: PV = pure vapor phase; CVT = chemical vapor transport; HTS = high-temperature solution; HS = hydrothermal solution; 10^2-10^4 represent the ratio $S = \lambda_0/h$, where λ_0 is the step separation and h the step height.

Source: Ref. 68, courtesy of North-Holland Publishing Co., Amsterdam.

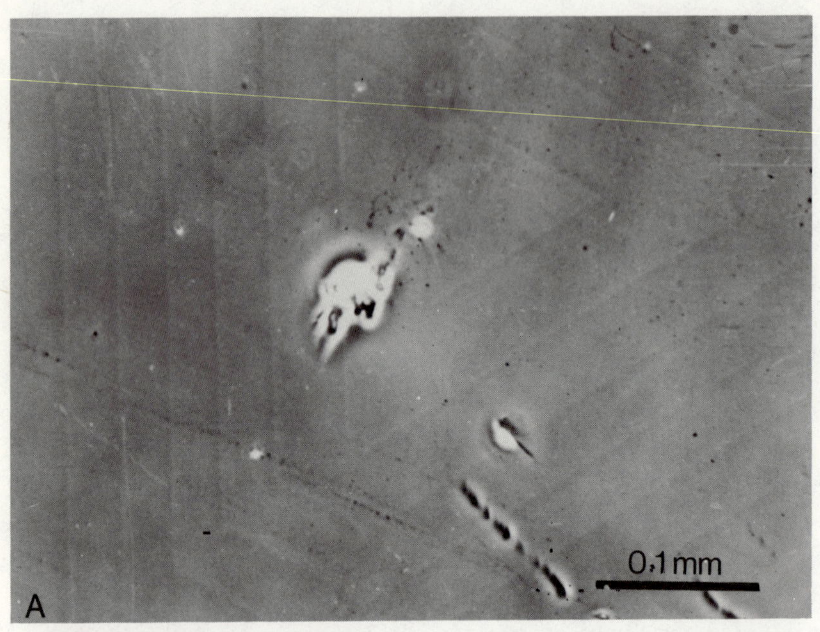
0,1 mm
A

20 µm
B

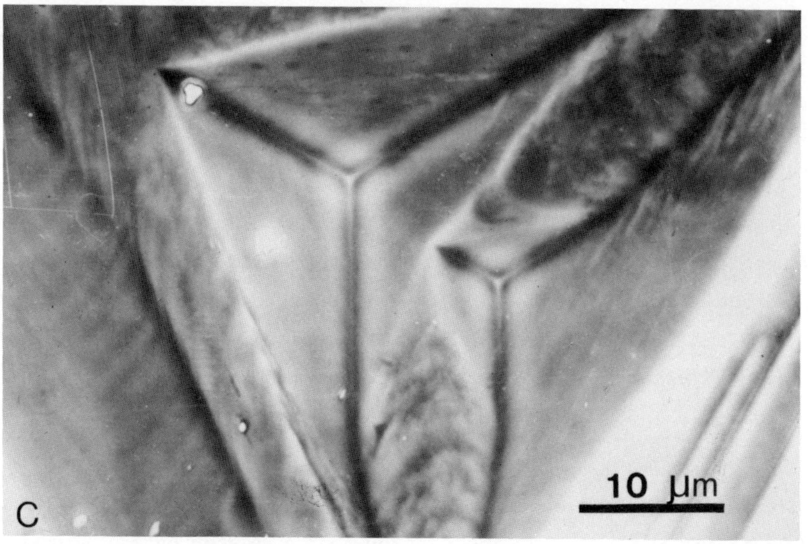

FIG. 65. A comparison of spirals observed on hematite crystals grown from vapor (A), (B) and solution phases (C): (A) natural chemical transport; (B) artificial chemical transport; (C) high temperature solution.

order of 10^3-10^4, whereas solution-grown crystals are 10^2-10^3 at the maximum. It is also noticed that solution-grown crystals show a general tendency to a more polygonized morphology than do vapor-grown crystals. This is clearly seen in Figs. 65 and 67.

2. Between PV and CVT: Both in SiC and phlogopite, crystals grown from a pure vapor phase show more circular spirals than those grown by CVT. The step separation is smaller in CVT than in PV, if only synthetic crystals are compared (SiC, Fig. 66). However, if a comparison is made between natural and synthetic crystals (phlogopite, Fig. 67), the former shows a wider step separation.
3. Between natural and synthetic crystals: In general, natural crystals exhibit more polygonal but irregular spirals and a one order of magnitude wider step separation than do synthetic crystals (hematite, phlogopite, beryl; Figs. 64, 65, and 67).
4. Other noticeable characteristics: Really circular spirals have been observed only on SiC crystals synthesized by the Lely method, in which the solid-fluid interaction energy should be very small. Other crystals investigated which have more complicated crystal structures than SiC show more polygonized forms. Figures 65, 66, and 67 demonstrate the above differences.

3. Discussion

We may assume that an essential difference between vapor and solution growth is the stronger ϕ_{sf} interaction in the latter case. Assuming that the temperature difference is not large if we compare vapor and solution growth, this means in agreement with Eq. (1a),

$$\alpha_{vapor} > \alpha_{solution}$$

and

$$\gamma_{vapor} > \gamma_{solution}$$

Since for a Kossel-like crystal $\gamma/KT \simeq \alpha/4$ (see Sec. II.B.7), if no surface diffusion is taken into consideration,

$$\lambda_{0\,vapor} > \lambda_{0\,solution}$$

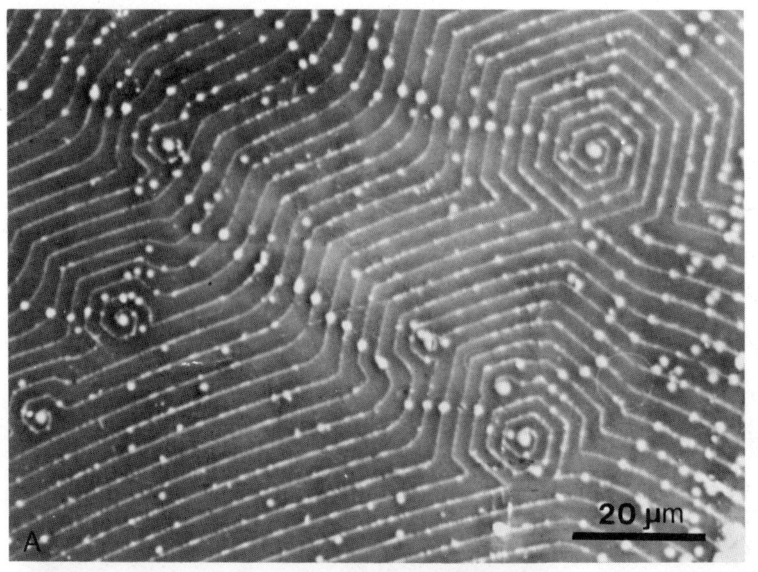

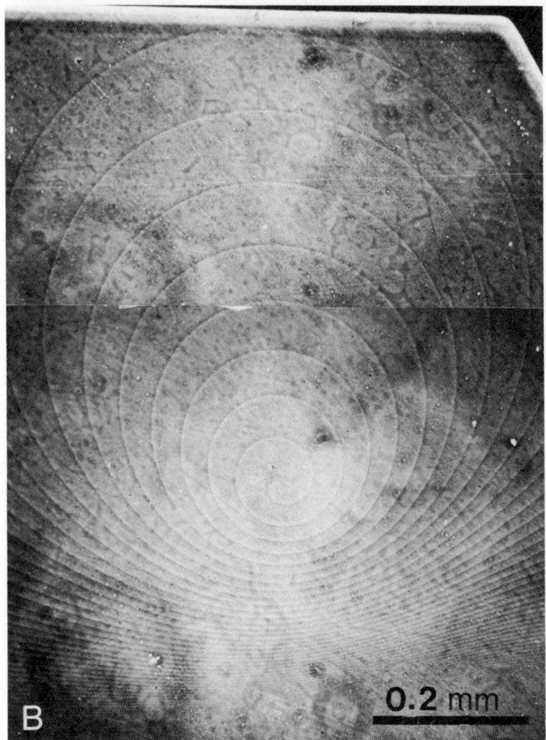

FIG. 66. A comparison of spirals observed on crystals grown (A) by chemical transport (Acheson method), and (B) from pure vapor phase (Lely method), photo by K. Narita; (0001), SiC.

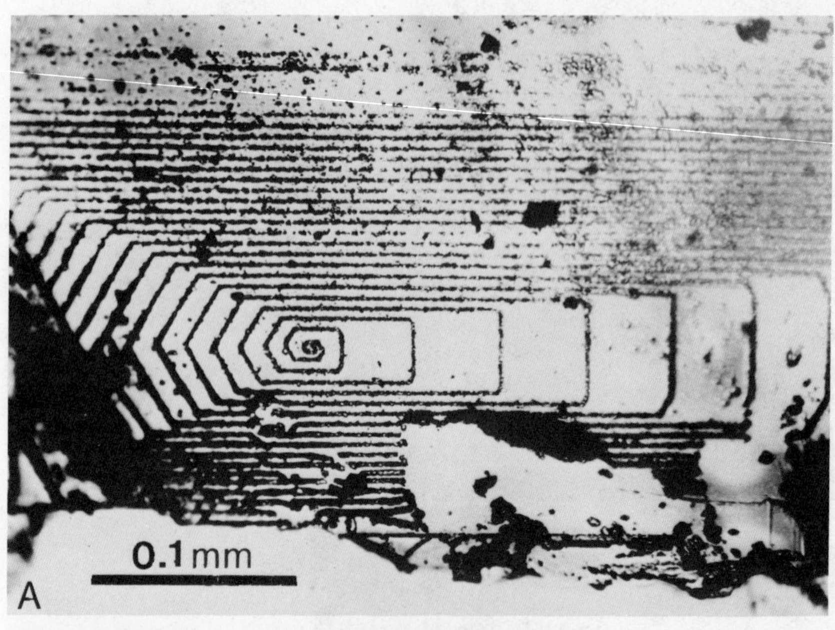
0.1 mm
A

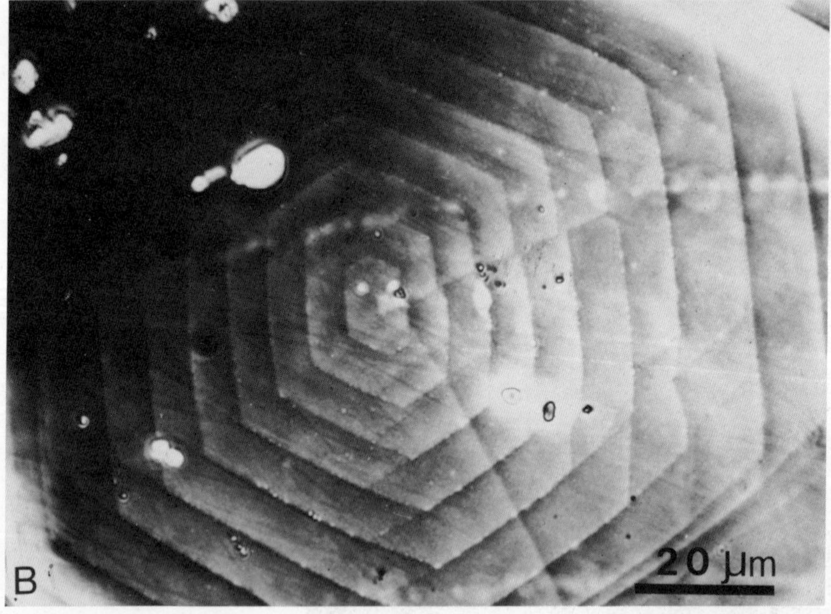
B
20 μm

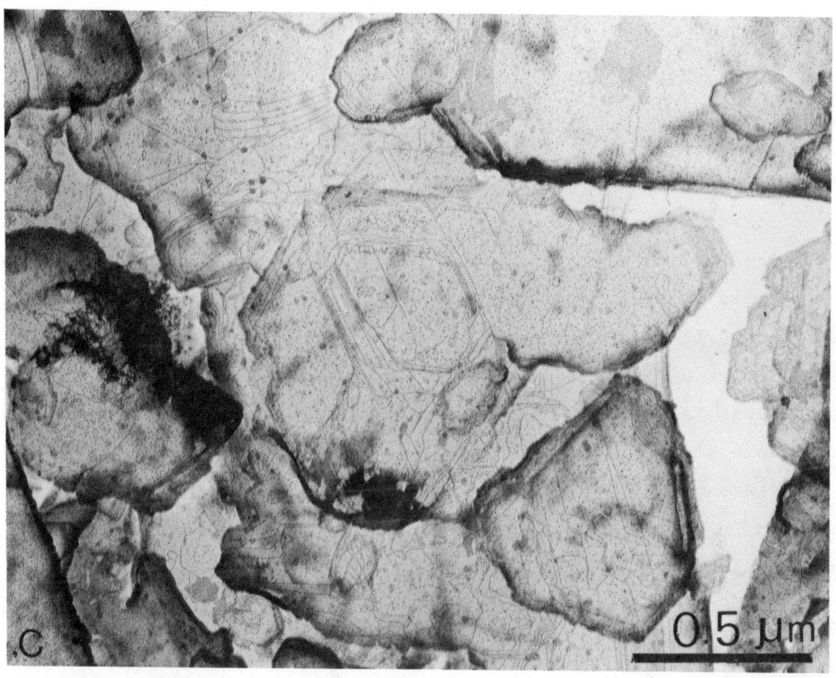

FIG. 67. A comparison of natural and artificial vapor phase growth and of vapor phase and solution phase growth, mica minerals: (A) natural phlogopite by chemical vapor transport,photo by S. Tomura, (B) synthetic fluor-phlogopite from pure vapor phase photo by Y. Endo, (C) natural muscovite (sericite) from solution phase photo. by M. Asakura.

since $\lambda_0 = 19r^*$ Eq. (4) and $r^* = \gamma\Omega/\Delta\mu$. We may assume that the mean displacement $\lambda_{s\ vapor} > \lambda_{s\ solution}$ (λ_s is the square of the displacement due to random walk of growth units over the surface, i.e., $\lambda_s = \sqrt{2D_s\tau_{deads}}$, where D_s is the surface diffusion constant and τ_{deads} the average life time of growth unit on the surface).

We may assume that

$$D_{s\ vapor} >> D_{s\ solution}$$

and

$$\tau_{deads\ vapor} > \tau_{deads\ solution}$$

which gives the above relation of λ_s.

A larger λ_s gives a larger λ_0, due to the back stress effect [48] as clearly formulated in the new theory of van der Eerden [49]. This makes the distance between spiral arms of vapor-grown crystals even larger as compared to the case of solution growth. A larger λ_s suppresses the tendency to polygonization. Hence we may expect that solution-grown crystals are more polygonized and have much smaller step distances.

In this argument we have neglected the fact that step integration may be also rate-determining.

On most crystals grown from HTS and HS, spirals show much narrower step separations than those grown from the vapor phase, though only occasionally are those having a fairly wide step separation observed on HTS crystals [69]. In such a case, CVT growth is usually suspected. This general tendency is in agreement with the theoretical expectation discussed above.

Crystals grown from a pure vapor phase show more circular spirals with a wider step separation compared to those grown from impure systems in which a stronger ϕ_{sf}, including an impurity effect, is involved. It is safe to conclude that the difference between SiC crystals synthesized by Acheson (impure) and Lely (pure) methods is due to this effect. In the case of natural phlogopite and synthetic fluor-phlogopite, the effect of T is added to the effect of ϕ_{sf}.

The difference between natural and synthetic crystals both in morphology and step separation (characteristic 3) may be attributed to a lower crystallization temperature, which increases α, and a lower supersaturation, which increases the step separation, in natural crystallization. The value of $\Delta\mu$ in natural vapor growth must have been very low compared to laboratory crystallization so as to compensate for the decrease of γ due to an impure environment. The wider step separation on natural crystals cannot be explained unless very low $\Delta\mu$ and low T are assumed, since a strong ϕ_{sf} is involved in natural crystallization. The polygonal but irregular spiral steps commonly observed on natural crystals are due to impurities and to foreign crystals.

The general characteristic 4 is probably due to the complicated structures of the investigated crystals and the ϕ_{sf} factor. Crystals with more complicated structures may have larger x_0 (distance between kinks) and hence show more polygonization. Although for a Kossel-like crystal, the transition from circular to polygonal far from the spiral center occurs at about $\alpha \simeq 7$, the transitional α value should be lower for actual crystals with more complicated structure.

From these, we may assume for ideal spirals with monomolecular step heights that

$$\alpha_{PV} > \alpha_{CVT} > \alpha_{HTS} \simeq \alpha_{HS}$$

$$\gamma_{PV} > \gamma_{CVT} >> \gamma_{HTS} \simeq \gamma_{HS}$$

$$\alpha_{natural} > \alpha_{laboratory}$$

$$\Delta\mu,\ T_{natural} < \Delta\mu,\ T_{laboratory}$$

These relations are in good accordance with the theroetical expectation discussed above and in Sec. II.B.12. In addition, it is worthwhile mentioning that the observed growth spirals on actual crystals are in surprisingly good accordance with computer-simulated spirals using the simplest Kossel-like crystal model.

I. Spirals with Macroscopic Step Heights

So far we have discussed growth spirals with steps of monomolecular or unit cell heights. Except in the case of higher-order polytypic crystals having large unit cell heights, the step heights of the observed growth spirals are on the order of 10 Å or less, and do not usually exceed 100 Å. When spiral layers with macroscopic step heights higher than a few hundred Å are observed, there is always evidence to suggest that they appear either due to bunching of monomolecular spiral steps, whose representative example is seen in Fig. 16, or occasionally due to twist boundaries, which consist of a grid of screw dislocations. Figure 68 is an example of the latter case, as observed on natural hematite. In all these cases, it is commonly observed that macrostep spirals occur together with spirals of monomolecular step heights. From this it is safe to assume that the unit step of growth spirals is monomolecular and thus crystallizing particles are ionic or atomic entities and not larger entities.

The above characteristic has been observed universally on all crystals grown from the vapor phase, both from a pure vapor and by chemical vapor transport. In contrast to this, most crystals grown from aqueous solutions do not show growth spirals having the above characteristics. Although it is true that the crystals grown from aqueous solutions may occasionally exhibit monomolecular growth spirals (e.g., Figs. 22, 48, and 59), the majority of the solution-grown crystals do not exhibit such spirals even if their surfaces are examined under the same conditions of observation as for vapor-grown crystals. Many natural crystals formed in hydrothermal veins (such as pyrite, sphalerite, quartz, barite, calcite, etc.) and also those grown in aqueous solution (like NaCl, KCl, $NaClO_3$) are examples in which monomolecular growth spirals have not been observed (e.g., see Ref. 89 for the case of pyrite). When growth spirals appear on these crystals, they always have macroscopic step heights, usually exceeding a few hundred Å. Spirals with monomolecular step heights are not observed on these surfaces, which makes a big contrast with the case of vapor-grown crystals where both monomolecular and bunched

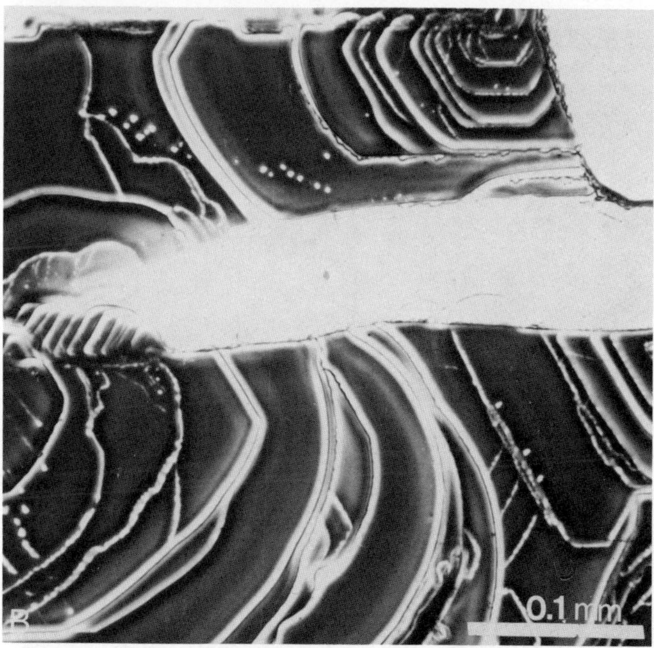

FIG. 68. Spirals with high steps originating from twist boundaries; (0001), hematite: (A) ordinary reflection photomicrograph; (B) phase contrast photomicrograph. White areas in (B) are portions inclined.

macrostep spirals coexist on the same surface. The macroscopic steps occurring on solution-grown crystals are always denticulated, and the surface of the spiral layers is wavy, not like atomistically flat surfaces commonly observed for monomolecular growth spirals. In Fig. 69, an example of such spirals is shown. Figure 27 is a representative of growth spirals commonly observed on synthetic diamonds.

Judging from σ^* in the R-σ relation (see Sec. I), dislocation controlled growth is expected at much lower supersaturations in the case of aqueous solution growth (σ^* is less than 2-3% [14]) than in the case of vapor growth (σ^* 25-50% [1]). The observations that ideal growth spirals are far less commonly observed on the crystals grown

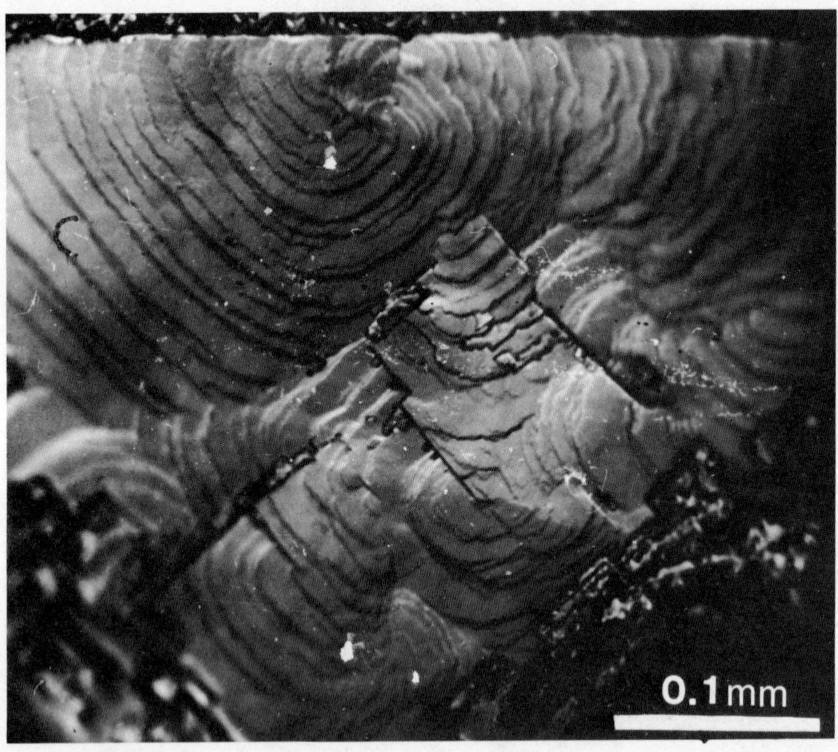

FIG. 69. Representative example of macrostep spirals observed on solution-grown crystals; (100), KCl. Photo by K. Tsukamoto.

from aqueous solutions than from vapor may partly be due to this difference. However, in order to account for the characteristics of macrostep spirals, i.e., high step heights, denticulated steps, and wavy surfaces, it is necessary to assume that the growth units must have been different between the two cases. As already discussed, ionic or atomic entities are assumed to be the growth units for ideal growth spirals, as well as for bunched macrostep growth spirals commonly observed on vapor-grown crystals, since both occur together on the same surface. In the case of macrostep spirals occasionally observed on some crystals grown from aqueous solution phase, this is not so. Monomolecular growth spirals are not associated with these macrostep growth spirals. To account for these characteristics, we have to assume larger entities as growth units in solution growth. If small particles are formed in the supersaturated solution and growth takes place by the accumulation of these particles, macrostep spirals having the observed characteristics may appear on solution-grown crystals. The existence of such small particles in supersaturated solutions has been suggested by Glasner [90,91], who assigned them the term *bloc nuclei*.

ACKNOWLEDGMENTS

This review presents a summary of the results of an international joint research project conducted by the groups of the two authors. It was financially supported by the Japan Society for the Promotion of Science. With the help of this support, the authors were able to visit one another. The authors also acknowledge the cooperation of the staffs and students in each group throughout this work and thank Prof. W. R. Wilcox for his critical reading of the manuscript and for correcting its English.

REFERENCES

1. W. K. Burton, N. Cabrera, and F. C. Frank, *Phil. Trans. Roy. Soc. London A243*, 299 (1951).
2. F. C. Frank, *Disc. Faraday Soc. 5*, 48 (1949).

3. L. J. Griffin, *Phil. Mag. 41*, 186 (1950).

4. A. R. Verma, *Crystal Growth and Dislocations*, Butterworths, London, 1953.

5. W. Dekeyser and S. Amelinckx, *Les Dislocation et la Croissance des Cristaux*, Masson, Paris, 1955.

6. S. Tolansky, *Surface Microtopography*, Longmans, London, 1960.

7. I. Sunagawa, *Amer. Mineral. 46*, 1216 (1961).

8. H. Bethge and K. W. Keller, *J. Cryst. Growth 20*, 105 (1972).

9. H. Bethge and K. W. Keller, *Optik 23*, 46 (1965).

10. K. W. Keller, *Growth of Crystals 7*, 145 (1969).

11. G. S. Gritsaenko and N. D. Samotoyin, Proc. Int. Clay Conf., Jerusalem, Israel, *3*, 391 (1966).

12. I. Sunagawa and Y. Koshino, *Amer. Mineral. 60*, 407 (1975).

13. R. Kaischev and E. Budevski, *Cont. Phys. 8*, 489 (1967).

14. T. Kuroda, T. Irisawa, and A. Ookawa, *J. Cryst. Growth 42*, 41 (1977).

15. I. Sunagawa, *J. Cryst. Growth 42*, 214 (1977); ibid. *45*, 3 (1978).

16. H. N. V. Temperley, *Proc. Cambridge Phil. Soc. 48*, 683 (1952).

17. H. J. Leamy, G. H. Gilmer, and K. A. Jackson, in *Surface Physics of Crystalline Materials* (J. M. Blakely, ed.), Academic Press, New York, 1975.

18. H. J. Leamy and G. H. Gilmer, *J. Cryst. Growth 24/25*, 499 (1974).

19. C. Van Leeuwen and F. H. Mischgofsky, *J. Appl. Phys. 46*, 1056 (1975).

20. C. Van Leeuwen, thesis, Technical University of Delft, 1977.

21. F. H. Mischgofsky, thesis, Technical University of Delft, 1977.

22. C. Van Leeuwen and P. Bennema, *Surf. Sci. 51*, 109 (1975).

23. G. H. Gilmer and P. Bennema, *J. Cryst. Growth, 13/14*, 148 (1972).

24. G. H. Gilmer and P. Bennema, *J. Appl. Phys. 43*, 1347 (1972).

25. S. W. H. de Haan, V. J. A. Meeussen, B. H. Veltman, P. Bennema, C. van Leeuwen, and G. H. Gilmer, *J. Cryst. Growth 24/25*, 491 (1974).

26. G. H. Gilmer, *J. Cryst. Growth 35*, 15 (1976).

27. J. P. van der Eerden, C. van Leeuwen, P. Bennema, W. L. van der Kruk and B. P. Th. Veltman, *J. Appl. Phys. 40*, 2124 (1977).

28. J. P. van der Eerden, *Phys. Rev. B13*, 4942 (1976).

29. G. H. Gilmer and K. A. Jackson, in *Crystal Growth and Materials* (E. Kaldis and H. J. Scheel, eds.), North Holland, Amsterdam, 1976, p. 79.

30. G. H. Gilmer, *J. Cryst. Growth 42*, 3 (1977).

31. H. Müller-Krumbhaar, in *Crystal Growth and Materials* (E. Kaldis and H. J. Scheel, eds.), North Holland, Amsterdam, 1976, p. 115.

32. R. H. Swendsen, *Phys. Rev. B15*, 5421 (1977).

33. H. J. F. Knops, *Phys. Rev. Lett. 39*, 766 (1977).

34. R. H. Swendsen, *Phys. Rev. B17*, 3710 (1978).

35. N. J. D. Knops and J. P. van der Eerden, *Phys. Letters 66A*, 77 (1978).

36. L. Onsager, *Phys. Rev. 65*, 117 (1944).

37. K. A. Jackson, in *Liquid Metals and Solidification*, Am. Soc. Metals, Cleveland, 1958, p. 174.

38. J. P. van der Eerden, P. Bennema, and T. A. Cherepanova, *Progr. Cryst. Growth Characterization 1*, 219 (1978).

39. H. van Beyeren, *Phys. Rev. Lett. 38*, 993 (1977).

40. P. Bennema and G. H. Gilmer, in *Crystal Growth: An Introduction* (P. Hartmen, ed.), North Holland, Amsterdam, 1973, p. 282.

41. P. Hartman, and W. G. Perdok, *Acta Cryst. 8*, 49, 521, 525 (1955).

42. P. Hartman and P. Bennema, *J. Cryst. Growth 49*, 145 (1980).

43. P. Bennema and J. P. van der Eerden, *J. Cryst. Growth 42*, 201 (1977).

44. N. Cabrera and M. M. Levine, *Phil. Mag. 1*, 450 (1956).

45. M. Ohara and R. C. Reid, *Modelling Crystal Growth Rates from Solution*, Prentice-Hall, Englewood Cliffs, N.J., 1973, p. 190.

46. H. Müller-Krumbhaar, T. W. Burkhardt, and D. Kroll, *J. Crystal Growth 38*, 13 (1977).

47. B. van der Hoek, J. P. van der Eerden, P. Bennema, and I. Sunagawa, *J. Cryst. Growth*, in press.

48. T. Surek, J. P. Hirth, and G. M. Pound, *J. Cryst. Growth 18*, 20 (1973).

49. J. P. van der Eerden, thesis, Catholic University of the Netherlands, Nijmegan, 1979.

50. R. Kaishev, *Growth Cryst. 3*, 29 (1962).

51. R. Boistelle and A. Doussoullin, *J. Cryst. Growth 36*, 336 (1976).

52. R. H. Swendsen, P. J. Kortman, D. P. Landau, H. Müller-Krumbhaar, *J. Cryst. Growth 35*, 73 (1976).

53. G. H. Gilmer, H. J. Leamy, K. A. Jackson, and H. Reiss, *J. Cryst. Growth 24/25*, 495 (1974).

54. N. Cabrera, *J. Chem. Phys. 21*, 1111 (1953).

55. B. van der Hock, P. Bennema, and J. P. van der Eerden, to be published.

56. I. Sunagawa and P. Bennema, *J. Cryst. Growth*, to be published.

57. I. Sunagawa and P. Bennema, *J. Cryst. Growth* 46, 000 (1979).

58. I. Sunagawa, K. Narita, P. Bennema, and B. van der Hoek, *J. Cryst. Growth 42*, 121 (1977).

59. N. Cabrera and D. A. Vermilyea, in *Growth and Perfection of Crystals* (R. H. Doremus, B. E. Roberts, and David Turnbull, eds.), Wiley, New York, 1958.

60. P. Bennema, J. Boon, C. van Leeuwen, and G. H. Gilmer, *Krist. Techn. 8*, 659 (1973).

61. A. H. Bennett, H. Jupnik, H. Osterberg, and O. W. Richards, *Phase Microscopy*, Wiley, New York, 1951.

62. M. Francon and S. Mallick, *Polarization Interferometers: Application in Microscopy and Macroscopy*, Wiley-Intersciences, New York, 1971.

63. S. Tolansky and V. G. Bhide, *J. Chim. Phys. 97*, 563 (1956).

64. S. Tolansky, *Multiple-Beam Interferometry*, Oxford University Press, London, 1948.

65. W. F. Knippenberg, *Philips Res. Rept. 18*, 161 (1963).

66. I. Sunagawa, *Amer. Mineral. 47*, 1139 (1962).

67. I. Sunagawa, *Sci. Rep., Tohoku Univ., Ser. 3, 12*, 239 (1974). (1974).

68. I. Sunagawa, *J. Cryst. Growth 45*, 3 (1978).

69. I. Sunagawa, *J. Cryst. Growth 1*, 102 (1967).

70. I. Sunagawa, in *Crystal Growth and Characterization* (R. Ueda and J. B. Mullin, eds.), North-Holland, Amsterdam, 1975, p. 347.

71. H. Komatsu and I. Sunagawa, *Amer. Mineral. 50*, 1046 (1965).

72. I. Sunagawa, IMA Symposium, Mineral. Soc. London, 63 (1968).

73. A. R. Verma, *Phil. Mag. 42*, 1005 (1951).

74. F. C. Frank, *Phil. Mag. 42*, 1014 (1951).

75. F. C. Frank, *Acta Cryst. 4*, 497 (1951).

76. A. R. Verma, *Z. Elektrochem. 56*, 268 (1952).

77. A. J. Forty, *Phil. Mag. 43*, 377 (1952).

78. I. Sunagawa, *Amer. Mineral. 45*, 566 (1960).

79. I. Sunagawa, *J. Mineral. Soc. Japan* (in Japanese) *12*, 117 (1975).

80. I. Sunagawa, *Lucas Mallada 7*, 45 (1960).

81. W. F. Berg, *Proc. Roy. Soc. A164*, 79 (1938).

82. F. C. Frank, in *Growth and Perfection of Crystals* (R. H. Dremus, B. W. Roberts, and D. Turnbull, eds.), Wiley, New York, 1958, p. 411.

83. P. Bennema and R. van Rosmalen, Proc. All Russian Conf., Erevan, Armenia, 1972; *Rost Kristallov 11*, 162 (1975).

84. R. Janssen-van Rosmalen and P. Bennema, *J. Cryst. Growth 32*, 293 (1976).

85. I. Sunagawa, *Mineral. J.* (Japan) *3*, 59 (1960).

86. I. Sunagawa, *Amer. Mineral. 47*, 1139 (1962).

87. I. Sunagawa, *Amer. Mineral. 49*, 785 (1964).

88. I. Sunagawa, *Forschr. Mineral. 52*, 515 (1975).

89. Y. Endo, *Bull. Geol. Surv. Japan, 29* 701 (1978).

90. A. Glasner and M. Tasa, *J. Cryst. Growth 13/14*, 441 (1972).

91. A. Glasner and M. Tasa, *Israel J. Chem. 12*, 799, 812 (1974).

92. H. Scholz, *Philips Techn. Rep. 28*, 316 (1967).

93. I. Sunagawa and P. J. Flanders, *Phil. Mag. 11*, 747 (1965).

94. K. Nassau, *J. Cryst. Growth, 13/14*, 12 (1972).

95. K. Watanabe, Y. Sumiyoshi, and I. Sunagawa, *J. Cryst. Growth 42*, 293 (1977).

96. I. Sunagawa, *Amer. Mineral. 49*, 1427 (1964).

97. I. Sunagawa and Y. Endo, *Mineral. Soc. Japan*, Spec. Pap. 1, 25 (1971).

98. I. Sunagawa, Y. Koshino, M. Asakura, and T. Yamamoto, *Forschr. Mineral. 52*, 217 (1975).

99. A. Baronnet, M. Amouric, and B. Chabot, *J. Cryst. Growth 32*, 37 (1976).

Chapter 2

SILICON NITRIDE

Donald R. Messier and William J. Croft

Army Materials and Mechanics Research Center
Watertown, Massachusetts

I. INTRODUCTION

During the last decade, the increasing importance of fuel conservation and interest in increased operating efficiency have stimulated renewed interest in the use of ceramic components in heat engines. Among the advantages of ceramics for such applications as compared to the metals currently used are their capability for operating at higher temperatures, the abundance of their raw materials, their strategic availability, and their potential low cost. Among the ceramic materials being considered for such applications, silicon nitride is a leading candidate.

The bulk of the literature on silicon nitride has appeared during the last two decades, but some work dates as far back as the last century. A German patent of 1895 by Mehner [1] contains one of the earliest references to silicon nitride. In this early work, the compound was synthesized by passing nitrogen over a mixture of carbon and silica heated in an electric furnace. Sinding-Larsen [2] later improved the process by reducing silica in a carbon arc furnace and reacting the resulting silicon vapor with nitrogen. Early patents noted that silicon nitride is readily produced by reduction of carbon-silica mixtures in nitrogen in the 1300-1450°C range using iron as a catalyst. The reason for limiting the temperature is to prevent the deposition of silicon carbide that occurs above 1450°C. This early work is summarized by Weiss and Engelhardt [3] who did a considerable amount of preparative chemistry and who recognized the composition of silicon nitride to be Si_3N_4. Another early summary of work on silicon nitride was given by L. Wöhler [4].

A renewal of interest in silicon nitride appears to have been stimulated by a paper of Collins and Gerby [5] who reported some of the physical and chemical properties of silicon nitride. In particular, they demonstrated its low coefficient of thermal expansion, a major reason for the current interest in silicon nitride. Their material was produced from powder compacts of elemental silicon that were heated at 1300°C in a nitrogen atmosphere. A more recent review of silicon nitride was published by Cutler and Croft [6], and a series of bibliographies on the same subject have been compiled by Messier and Murphy [7].

At present, most of the silicon nitride being used is produced by two techniques, namely, by direct reaction of the elements and by chemical vapor-phase reactions. Both techniques may be used to form either silicon nitride powder or relatively dense, free-standing shapes. Materials for structural applications are generally made by direct reaction or by consolidation of silicon nitride powder with heat and often with applied pressure. Thin insulating films of more or less pure silicon nitride for electronic applications have been widely produced by chemical vapor deposition [8].

Some work has also been done on the preparation of silicon nitride by thermal decomposition of various metal-organic precursors. An extensive study in this area was reported by M. Billy [9] who, in the same paper, gives a comprehensive review of earlier work on silicon nitride.

II. CRYSTALLOGRAPHY

The relationship between the two forms of silicon nitride (α and β) has been the subject of much controversy. The selected lattice parameter data in Table 1 for the two phases emphasize this point; there exist considerable differences in reported values, particularly for $\alpha\text{-}Si_3N_4$.

The existence of the two forms was first established by Turkdogan et al. [10]. Although alternative interpretations of their powder x-ray diffraction data existed, Hardie and Jack [11] established

TABLE 1

Unit Cell Data for Silicon Nitride (Dimensions in Å)

α Phase		β Phase		
a_o	c_0	a_0	c_0	Ref.
7.748 ± 0.001	5.617 ± 0.001	7.608 ± 0.001	2.9107 ± 0.0005	Hardie and Jack [11]
7.753 ± 0.004	5.618 ± 0.004	7.606 ± 0.003	2.909 ± 0.002	Ruddelesden and Popper [13]
7.758 ± 0.005	5.623 ± 0.005	7.603 ± 0.005	2.909 ± 0.003	Forgeng and Decker [24]
7.76 ± 0.05	5.64 ± 0.006	7.59 ± 0.006	2.92 ± 0.008	Narita and Mori [16]
7.752 ± 0.003	5.619 ± 0.001	7.604 ± 0.001	2.907 ± 0.001	Suzuki [25]
7.755 ± 0.005	5.616 ± 0.005	7.606 ± 0.005	2.907 ± 0.003	Thompson and Pratt [22]
7.7520 ± 0.0007 (wool)	5.6198 ± 0.0005	7.608 ± 0.005	2.911 ± 0.001	Wild et al. [14]
7.7533 ± 0.0008 (Needles)	5.6167 ± 0.0006			Wild et al. [24]
7.765 ± 0.001	5.622 ± 0.001			Marchand et al. [17]
7.753 ± 0.001 (Black)	5.625 ± 0.001			Galasso et al. [26]
7.753 ± 0.001 (White)	5.618 ± 0.001			Galasso et al. [26]
7.760 ± 0.001	5.613 0.001			Priest et al. [20]
7.766 ± 0.010	5.615 0.008			Kohatsu and McCauley [18]
7.818 ± 0.003	5.591 0.004			Kato et al. [19]

that both forms are hexagonal with the c dimension of α-Si_3N_4 being approximately twice that of β-Si_3N_4.

Hardie and Jack [11] reported that β-Si_3N_4 had a phenacite-type structure as had been suggested by Juza and Hahn [12] for the analogous Ge_3N_4. The original work of Hardie and Jack was confirmed by Ruddlesden and Popper [13], and by Wild et al. [14]. Comparison between the β structure, having a unit cell comprising Si_6N_8, and phenacite (Be_2SiO_4) shows that the former has Si on both the Be and Si sites in the latter structure. The O sites are occupied by N and the result is a β-Si_3N_4 structure in which each Si is at the center of an irregular tetrahedron of N atoms. The tetrahedra share corners in such a way that each N is common to three tetrahedra. The symmetry of β-Si_3N_4 is hexagonal and Hardie and Jack [11] assigned it to the space group $P6_3/m$ with 2N in (c) at $\pm(1/3, 2/3, 1/4)$; 6N in (h) at $\pm(x, y, 1/4)$, $\pm(y, x-y, 1/4)$, $\pm(y-x, x, 1/4)$; and with 6-Si also in sites (h). Atomic parameters were reported by Hardie and Jack as $X_N = 0.333$, $Y_N = 0.333$, $X_{Si} = 0.172$, and $Y_{Si} = 0.213$. The estimated precision of these parameters was ± 0.005. Wild et al. [14] report the bond lengths as follows: Si-N_1 1.730 and 1.739, Si-N_2 1.745 Å. Although a more recent β-Si_3N_4 structure determination by Borgen and Seip [15] disagrees slightly with that of Wild et al. [14], the R factor of 0.021 given by the latter group is considerably less than the rather high 0.109 reported by Borgen and Seip. In any event, it may be concluded that all investigators are in basic agreement as to the structure of β-Si_3N_4.

Until recently, considerable disagreement has existed regarding the nature of the structure of α-Si_3N_4. Early structure determinations by Hardie and Jack [11], Ruddlesden and Popper [13], Narita and Mori [16], and Marchand et al. [17] seemed in agreement, but later, Wild et al. [14] proposed that α-Si_3N_4 was actually an oxynitride with O replacing N in some sites and with other N sites vacant. This structure was found to be consistent with their structure determination from powder data, and they offered supporting evidence from preparative chemistry, oxygen analyses, and density measurements. Wild et al. therefore proposed that, rather than being considered low- and high-temperature polymorphs, α-Si_3N_4 and β-Si_3N_4 should be regarded respectively as "high-oxygen-potential" and "low-oxygen-potential" modifications. They further suggested composition limits for α-Si_3N_4 of between $Si^{+4}_{11.5}N^{-3}_{15}O^{-2}_{0.5}$ and $Si^{+4}_{10}Si^{+3}_{2}N^{-3}_{15}O^{-2}_{0.5}$. The former composition corresponds to an oxygen content of 1.48 w/o and the latter to 0.90 w/o. Wild et al. refined the structure by use of site occupancy values to an R factor of 3.9%, unusually low for a structure based on powder data.

Evidence in the more recent literature is overwhelmingly against the proposed oxynitride formula for α-Si_3N_4. Structure determinations on α-Si_3N_4 single crystals by Kohatsu and McCauley [18] and Kato et al. [19] agree with the one by Marchand et al. [17], but not with that of Wild et al. [14]. In none of three former studies was it necessary to assume that α-silicon nitride comprised anything but Si_3N_4, and, furthermore, the determination of Kato et al. [19] was done on an α crystal with an analytically determined oxygen content of only 0.05 w/o. Other chemical analyses on α-Si_3N_4 materials have also given oxygen contents considerably less than that required by the oxynitride formula; Priest et al. [20] found 0.30 w/o oxygen in α-Si_3N_4 prepared by CVD, and Edwards et al. [21] found oxygen contents in reaction-bonded Si_3N_4 that were less than half the amount required by that formula. It is therefore concluded, as was done by Kohatsu and McCauley [18], that although oxygen might substitute for nitrogen in α-Si_3N_4, there is no structural requirement for its presence.

In accordance with the above discussion, we accept the structure determinations for α-Si_3N_4 from powder data by Ruddlesden and Popper [13] and from single-crystal data by Marchand et al. [17] (later confirmed by Kohatsu and McCauley and by Kato et al.) as being correct. The symmetry of α-Si_3N_4 is trigonal, its space group is P3/c, and its unit cell comprises $Si_{12}N_{16}$. Tables 2 and 3 include the atomic coordinates and bond lengths for α-Si_3N_4. As seen in Table 1, Kato et al. [19] give unit cell dimensions for α-Si_3N_4 that differ noticeably from those reported by previous workers. They attribute this to

TABLE 2

Atomic Coordinates of α-Si_3N_4

	Si_1	Si_2	N_1	N_2	N_3	N_4
Equipoint	6c	6c	6c	6c	2b	2a
x	0.5130(6)[a]	0.1680(7)	0.6124(20)	0.3199(18)	2/3	0
y	0.4305(6)	0.9150(6)	0.9592(20)	0.0045(20)	1/3	0
z	0.658(9)	0.4504(9)	0.4343(26)	0.7045(32)	0.6015(36)	0.4520(45)

[a]Standard deviation of last significant figure in parentheses.

Source: Ref. 18.

possible differences in purity between their specimen and others, but this explanation is highly unlikely. The difference could be a result of some peculiarity in their measurement technique, but insufficient information exists to speculate further on this point.

Thompson and Pratt [22] studied the crystal structures of α- and β-Si_3N_4 by constructing models of each based on the data of

TABLE 3

Tetrahedral Bond Lengths for α-Si_3N_4

	Si_1	$-N_1$	1.708(15)[a] Å
		N_2	1.775(15)
		N_2	1.738(13)
		N_3	1.730(6)
Mean	Si_1	-N	1.738
	Si_2	$-N_1$	1.763(16)
		N_2	1.758(40)
		N_2	1.700(40)
		N_4	1.731(5)
Mean	Si_2	-N	1.738

[a]Standard deviation of last significant figure in parentheses.

Source: Ref. 18.

Ruddlesden and Popper [13]. They were particularly interested in the distortion of the nitrogen tetrahedra necessary to form stable structures. The c/a ratios determined from their models agreed well with those determined from x-ray diffraction data for both structures. Taking into account the partial ionicity of the Si-N bond, Thompson and Pratt calculated the interatomic separation of Si and N to be 1.77 Å. This value was in good agreement with those determined for β-Si_3N_4 from their model (1.79 Å), and with the experimental values of 1.72 and 1.75 Å found by Hardie and Jack [11].

Figure 1 (Thompson and Pratt [22]) is a projection of the basal plane common to each of the Si_3N_4 structures. Note the large central void surrounded by a hexagonal array of atoms; a cylinder of radius 1.5 Å can be accommodated in it. The smaller, six-sided space seen in the figure contains a void of 0.9 Å radius.

Another illustration of the structures of α- and β-Si_3N_4, after Messier [23], appears in Fig. 2. The structure was described similarly by Hardie and Jack [11]. As shown in Fig. 2, both structures can be formed completely from basic units comprising puckered rings, each containing four Si and four N atoms. Six of these rings join to form a sheet containing the large central void whose dimensions

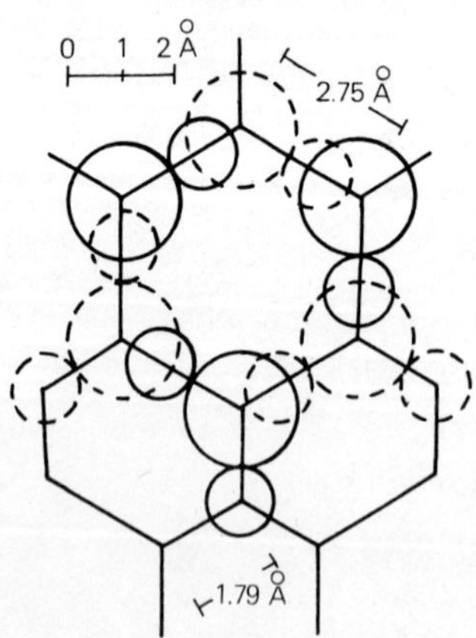

FIG. 1. Basal plane common to alpha and beta silicon nitride. (From Ref. 22.)

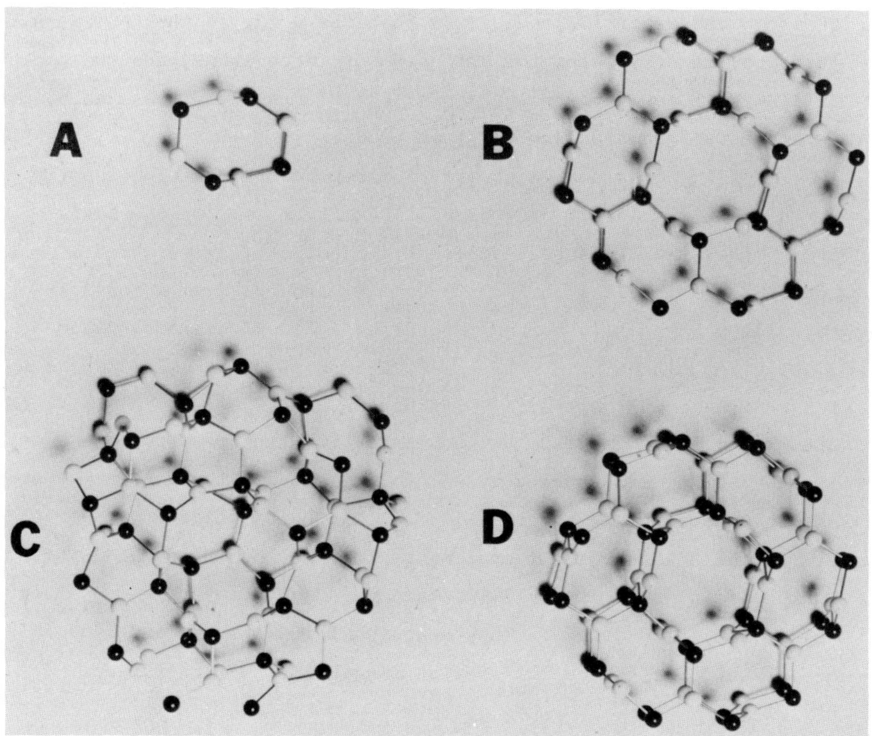

FIG. 2. Models illustrating the crystal structure of silicon nitride. The dark balls represent silicon and the light ones nitrogen. (A) Eight-membered puckered ring; (B) six puckered rings joined to form a basal plane sheet common to both structures; (C) alpha silicon nitride; (D) beta silicon nitride. (From Ref. 23.)

are given in Fig. 1. The basal plane sheets are sufficient to construct the entire crystal structures of both α- and β-Si_3N_4.

As seen in Fig. 2, the β-Si_3N_4 structure comprises basal plane sheets stacked in the sequence A, A, A, etc. The β-Si_3N_4 structure is relatively unstrained and contains large channels. On the other hand, in the α-Si_3N_4 structure the sheets are stacked alternately right side up and upside down in the sequence A, B, A, B, etc. The α-Si_3N_4 structure is somewhat strained as compared to that of β-Si_3N_4. Because of these stacking differences, the positions where voids exist in the latter structure contain large pockets in the α structure.

Thompson and Pratt [22] suggested that the voids in the β-Si_3N_4 structure provide easy paths for diffusion of large atoms, and the pockets in α-Si_3N_4 traps for similar atoms; while this seems reasonable, no experimental proof exists for such behavior.

In 1910 Weiss and Engelhardt [3] reported a measured density for silicon nitride of 3.44 g cm^{-3} (3440 kg m^{-3}). On the basis of their x-ray measurements Ruddlesden and Popper [13] reported a density for α-Si_3N_4 of 3.18 g cm^{-3} (3180 kg m^{-3}) and for β-Si_3N_4 of 3.19 g cm^{-3} (3190 kg m^{-3}). Thompson and Pratt [22] calculated a value of 3.185 g cm^{-3} (3185 kg m^{-3}) for α-Si_3N_4 and 3.196 g cm^{-3} (3196 kg m^{-3}) for β-Si_3N_4 from their x-ray diffraction data. Wild et al. [14] in attempting to demonstrate the necessity for the presence of oxygen in the α-Si_3N_4 structure very carefully measured the density of small single crystals. In the case of β-Si_3N_4 the calculated value was 3.192 g cm^{-2} and by the flotation method they measured 3.192 ± 0.002. On the basis of their oxygen-containing structure for α-Si_3N_4 they calculated 3.168 g cm^{-3}. They measured 3.169 ± 0.004 and used this agreement to reinforce the idea of oxygen in the structure.

III. THERMODYNAMICS AND PHASE EQUILIBRIA

A. Introduction

Any analysis of thermodynamic measurements and phase diagram determinations has to consider the somewhat controversial topic of the nature of α- and β-silicon nitride. As mentioned in the previous section, it now seems clear that earlier suggestions [14,27-31] that α-Si_3N_4 is actually an oxynitride were wrong. Although it is still unresolved as to whether or not one of the forms may be metastable or perhaps stabilized by impurities, it must be concluded from available evidence that both α- and β-silicon nitride are stoichiometric compounds represented by the formula Si_3N_4.

Existing thermodynamic data on condensed Si_3N_4 were for the most part obtained from measurements on materials that consisted of mixtures of the two phases in undetermined amounts. While it would be desirable to have measurements on phase-pure materials, it appears that differences between the thermodynamic properties of α- and β-Si_3N_4

could well be smaller than the error limits of customary measurement techniques; as already discussed in the previous section, α- and β-Si_3N_4 have similar crystal structures and densities, and first-order bonding is the same in each phase.

Despite the fact that α-Si_3N_4 cannot be considered to be an oxynitride, the importance of oxygen in the chemistry of the silicon-nitrogen system cannot be overestimated. The majority of the materials investigated to date, whether on a scientific or engineering basis, have been contaminated to some extent with oxygen. Oxygen impurity levels range from 0.1 w/o or so, to, more typically, 1 w/o or even greater. Possible effects of this impurity were unconsidered, and probably unrecognized in early thermodynamic measurements [32,33]. In any practical system oxygen is present to some extent, and the phase diagrams reviewed in this section all recognize this fact.

B. Thermodynamic Data

The first significant measurements of the silicon-nitrogen equilibrium were made by Hincke and Brantley by a mamometric technique in the temperature range from 1606 to 1802 K [32]. More extensive measurements made by Pehlke and Elliot [33] by a similar technique over the temperature range from 1673 to 1973 K agreed well with the earlier ones. The data for "α-Si_3N_4" in the JANAF tables [34] and in Table 4 are based on the results of these two sets of measurements. Gibbs free energy data obtained from equilibration of β-Si_3N_4 with Si-Fe alloys by Blegen [35] are in excellent agreement with those given in the JANAF tables [34] nominally for "α-Si_3N_4."

The data given in Table 4 for the species SiN(g) and Si_2N(g) were calculated from data from various sources [34].

The Gibbs free energy data listed for silicon oxynitride in Table 4 were derived from effusion measurements by Blegen [35]. Insufficient data are available for calculation of the other quantities listed in Table 4, or for calculation of room temperature Gibbs free energy data.

TABLE 4

Thermodynamic Data for Compounds in the Si-O-N System

Species	C°_p, J $mol^{-1} deg^{-1}$	S°, J $mol^{-1} deg^{-1}$	$\Delta H^\circ_{f298.15\ K}$, kJ mol^{-1}	$\Delta G^\circ_{f298.1\ K}$, kJ mol^{-1}	Ref.
α-Si_3N_4(s)	99.5	113.0	-744.8	-647.4	[34]
SiN(g)	30.2	216.7	372.4	341.7	[34]
Si_2N(g)	49.9	256.4	397.5	360.8	[34]
Si_2ON_2(s)[a]	--	--	--	-420 (1800 K)	[35]

[a]Data on this compound are at present insufficient for determination of other thermodynamic parameters.

C. Thermal Decomposition of Si_3N_4

The results given in this section are from the thermodynamic analysis by Singhal [36] whose calculations were made using data from sources already quoted herein [32-35]. Singhal chose to ignore the Gibbs free energy data of Wild et al. [29] and Colquhoun et al. [30] because of poor agreement with the other investigators, and because of suspected experimental inadequacies [36,37].

Silicon nitride dissociates into Si (solid or liquid depending upon the temperature) according to the reaction [38]:

$$Si_3N_4(s) = 3Si(s,l) + 2N_2(g) \quad (1)$$

Figure 3 from Singhal [36], shows the dissociation pressure of $N_2(g)$ over $Si_3N_4(s)$ as a function of temperature. The temperature at which the decomposition pressure reaches 1 atm (101.3 kPa) is 2151 K.

The vaporization behavior of Si_3N_4 is complex and it is still incompletely characterized with respect to the nature of the vapor species [39]. Possible reactions involved in the vaporization of Si_3N_4 include:

$$2Si_3N_4(s) \rightarrow 6SiN(g) + N_2(g) \quad (2)$$

$$4Si_3N_4(s) \rightarrow 6Si_2N(g) + 5N_2(g) \quad (3)$$

$$Si_3N_4(s) \rightarrow 3Si(g) + 2N_2(g) \quad (4)$$

$$2Si_3N_4(s) \rightarrow 3Si_2(g) + 4N_2(g) \quad (5)$$

$$Si_3N_4(s) \rightarrow Si_3(g) + 2N_2(g) \quad (6)$$

Figure 4 (after Singhal [36]) shows the partial pressures of the various vapor species in reactions (2) to (6) as functions of temperature in N_2 at a pressure of 1 atm. (101.3 kPa). It is apparent that the predominant vapor species over Si_3N_4 are SiN(g) and Si(g); it should be noted, however, that at the relatively high temperature of 1600 K, the partial pressures of both species are still extremely small, 4×10^{-10} atm. (4×10^{-5} Pa). As noted by Singhal, higher nitrogen pressures reduce the decomposition of $Si_3N_4(s)$ even further.

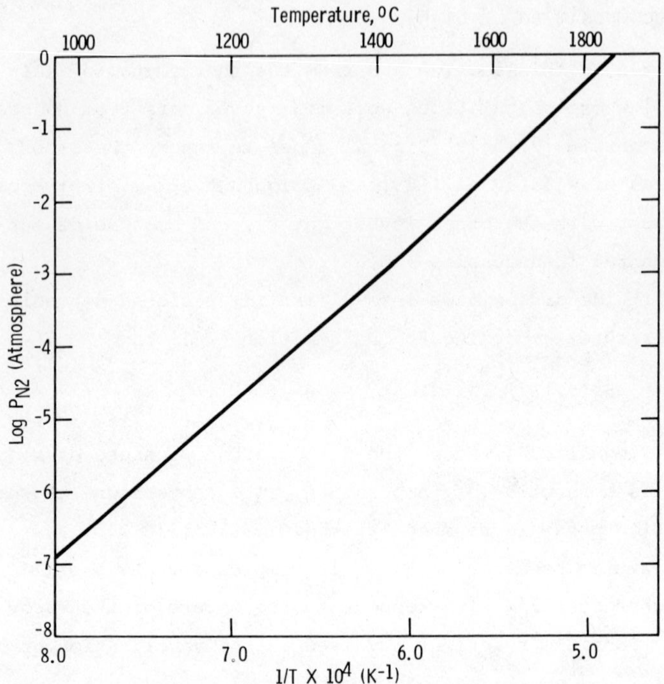

FIG. 3. Dissociation pressure of nitrogen gas over solid silicon nitride. (From Ref. 36.)

D. The System Si-N-O

The first phase diagram data for the system Si-N-O were based on the assumption that α-Si_3N_4 is in fact an oxynitride [14,27-30]. As already mentioned, the bulk of the evidence now in the literature indicates that such is not the case. Critical reviews of existing thermodynamic data by Singhal [36] and Blegen [35] support this conclusion. Those two workers devised equilibrium diagrams for the system Si-N-O based on still questionable assumption that β-Si_3N_4 is the only stable form of silicon nitride.

Figure 5 is the calculated equilibrium diagram for the system Si-N-O at 1800 K as given by Blegen [35]. The diagram of Singhal agrees well with this one. A remarkable conclusion from this diagram, as well as earlier ones, is that Si_3N_4 is stable only at oxygen

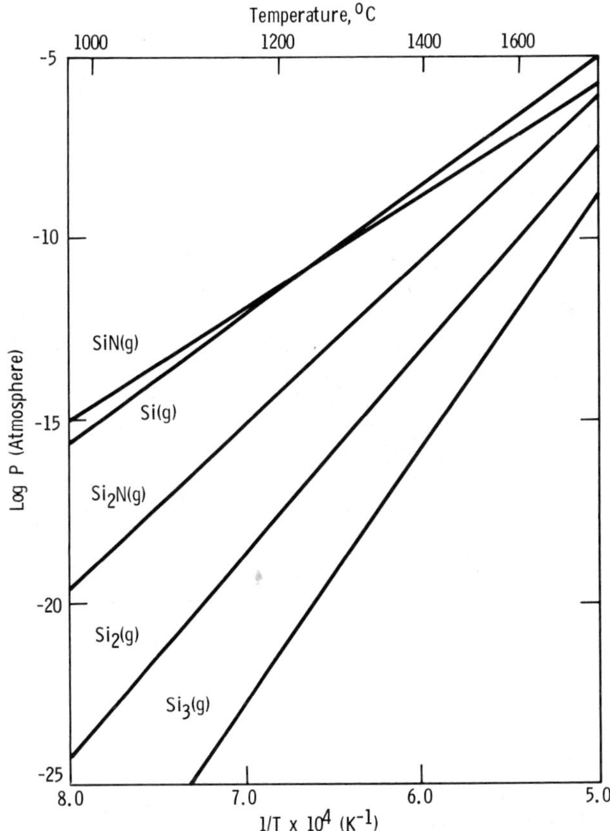

FIG. 4. Partial pressures of various volatile species over solid silicon nitride in nitrogen at 1 atm. (From Ref. 36.)

partial pressures far lower than one would expect to find in a typical experimental system. There is no reason to believe that the thermodynamic data are grossly in error, and the relative ease of preparation of Si_3N_4 has been amply demonstrated; it must therefore be concluded that kinetic factors are overwhelmingly predominant in the synthesis of that compound.

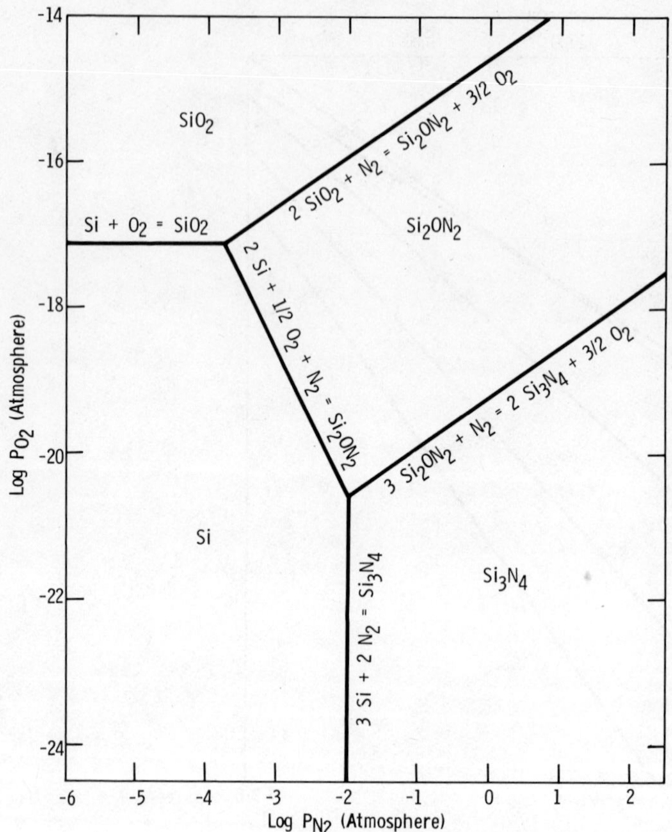

FIG. 5. Calculated equilibrium diagram for the system silicon-oxygen-nitrogen at 1800 K. (From Ref. 35.)

IV. FORMATION AND PHASE TRANSFORMATIONS

A. Introduction

A substantial volume of literature has been generated, particularly since the late 1960s, on kinetics and mechanisms of chemical reactions involved in the formation of Si_3N_4. A critical review of that literature, however, shows that considerable uncertainty still exists regarding the exact nature of those reactions. It has become apparent that the processes involved are considerably more complex than could have been appreciated by early investigators, and that a large effort was necessary to merely pose the critical questions that had to be answered.

An excellent review on work that has been done on silicon-nitrogen reactions with emphasis on the fabrication of reaction-bonded Si_3N_4 has been published recently by Moulson [40]. Considerable work in this area has been accomplished by him and his colleagues at the University of Leeds and his review is clearly the most comprehensive to date on the subject. Another recent review on nitridation and reaction bonding was published by Riley [41], and an excellent review of work prior to 1959 has been given by Billy [9].

B. Kinetics of Silicon-Nitrogen Reactions

Much of the research that has been conducted on silicon-nitrogen reactions was stimulated by the interest in Si_3N_4 as a structural material, particularly in the form of reaction-bonded silicon nitride (RBSN). That material is formed by heat-treating shapes compacted from Si powder in nitrogenous atmospheres over various heating schedules in the temperature range from 1100 to 1400°C. During the process, the Si is more or less completely converted to Si_3N_4 and the result is a relatively dense, strong body of that material. Additional information on the reaction bonding process appears in Sec. V of this chapter.

Quantitative conversion of Si to Si_3N_4 would be accompanied by a gravimetrically determined weight increase of approximately 66.7%. However, observed weight gains upon complete conversion are invariably less than this amount, and Billy [9] appears to be the first to recognize that this discrepancy is a result of material lost from the reacting mass by vapor-phase transport. This circumstance, in hindsight, was an early sign of potential complications in evaluating reaction kinetics.

A paper published by Popper and Ruddlesden [42] in 1961 would have been, as suggested by Moulson [40], "a firm foundation on which subsequent research should have been based." Popper and Ruddlesden described qualitatively a number of features that would be of critical importance to future understanding and control of the reaction process. They recognized that the exothermicity of the reaction could cause melting of unreacted silicon unless the heating rate was carefully controlled, that impurities in the nitriding atmosphere may affect the nitriding rate, and that iron, calcium fluoride, and possibly impurities in commercial Si act as catalysts for the reaction. They also raised the possibility that a silica surface layer might retard the reaction, and reported as did Billy [9] that observed weight gains upon nitridation were typically 61% vs. the theoretical value of 66.7%. Popper and Ruddlesden were unsuccessful in obtaining reproducible data on the kinetics of the silicon-nitrogen reaction, and concluded that the variability in their experiments was due to some uncontrolled factor that was equally as important as temperature.

A later paper by Messier and Wong [43] summarized previous work [9,25,42,44-48] on the kinetics of the silicon-nitrogen reaction. They pointed out that variability such as that observed by Ruddlesden and Popper in their own experiments are even more pronounced when comparing the results of various investigators. For example, the reaction kinetics had been described as logarithmic [9,22], approximately parabolic [42], and linear [47]. Messier and Wong [43] and Messier et al. [48] presented the first convincing evidence of the effect of small oxygen partial pressures on nitridation, and, in the former paper, suggested that the "active" and "passive" oxidation suggested by Wagner [49] played a significant role in the nitridation process. They also suggested that iron impurities in the Si result in the formation of iron silicide liquids that affect the process, and gave evidence for the formation of α-Si_3N_4 whiskers via the VLS [50] mechanism proposed by Gribkov et al. [51].

A substantial portion of the more recent work on the nitridation of Si has been conducted at the University of Leeds by Moulson and Riley and various coworkers [52-64]. This work has been nicely summarized by Moulson [40] in his recent review paper. Whereas much of the previous work on the subject had been more or less applications-oriented, the Leeds researchers have made a concerted effort to develop an understanding of the basic mechanisms of the nitridation of Si. Their emphasis has been on working with high-purity materials and on careful control of experimental variables. The summary of the Leeds work given here draws heavily from Moulson's [40] paper, and the interested reader is urged to consult the original references [52-64] for additional detail.

The principal chemical reactions involved in the nitridation of Si are as follows:

$$3Si(s) + 2N_2(g) = Si_3N_4(s) \quad (7)$$

$$3Si(g) + 2N_2(g) = Si_3N_4(s) \quad (8)$$

$$Si(s) + SiO_2(s) = 2SiO(g) \quad (9)$$

Reaction (9) is included because of its role in causing loss of Si from the system, and because, under certain conditions, SiO(g) may participate in Si_3N_4-forming reactions.

The model developed by Atkinson et al. [56-58] for the nitridation of Si is illustrated in Fig. 6. This model resulted from careful observations of the nitridation of Si at various temperatures and N_2 pressures. Shown in the figure are the nucleation and growth processes responsible for the formation of monolithic Si_3N_4 on Si as well as evaporation and surface diffusion processes that may contribute to the growth of this layer. Also illustrated is another mechanism of significance: the formation by vapor-phase reactions of Si_3N_4 in the whisker-type morphology often observed as a product of nitridation.

The model shown in Fig. 6 is consistent with many observations of reaction kinetics: during the early stages of nucleation and growth, linear kinetics are observed; as the nitrided layer extends over the surface, the supply of Si to the reaction sites decreases, as does the reaction rate; when the nitride film covers the surface completely, the rate falls to nearly zero. The model shown in Fig. 6 is also qualitatively consistent with observations on the dependence of reaction kinetics on temperature and N_2 pressure. In a further study of reaction kinetics, Longland and Moulson [62] concluded as did Blegen [35] that α-Si_3N_4 is formed by vapor-phase reactions whereas β-Si_3N_4 forms via direct nitridation of the Si surface.

It has long been established that Fe has a significant influence on the kinetics of nitridation of Si [42], and Boyer and Moulson [60,63]

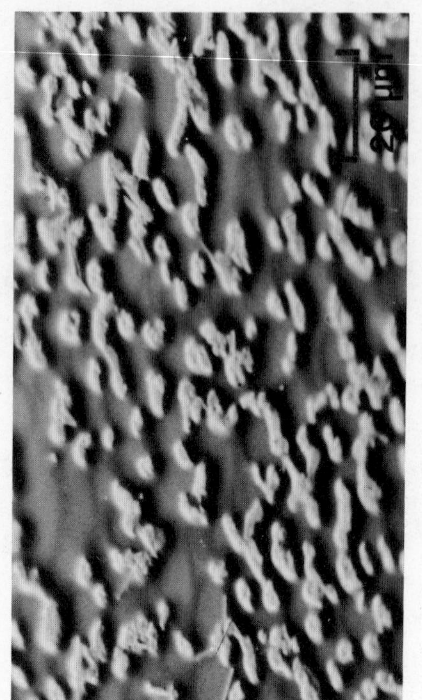
20 µm

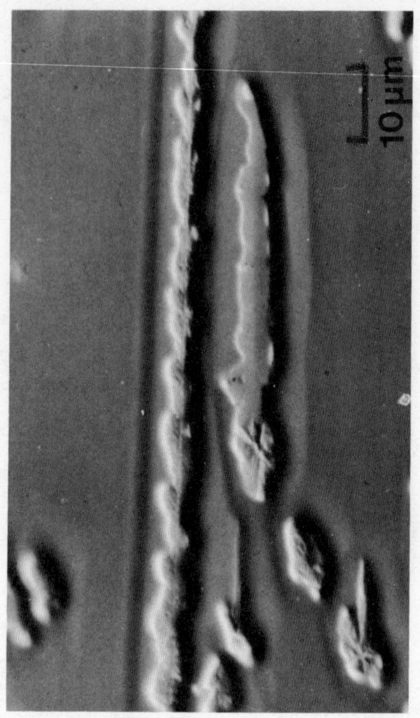
10 µm

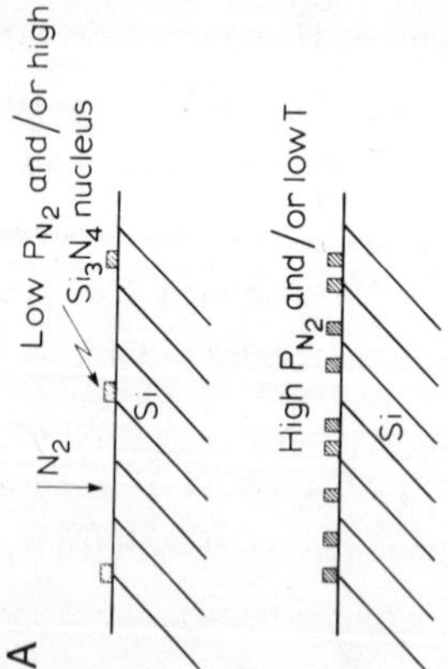
A
N_2
Low P_{N_2} and/or high T
Si_3N_4 nucleus
Si
High P_{N_2} and/or low T
Si

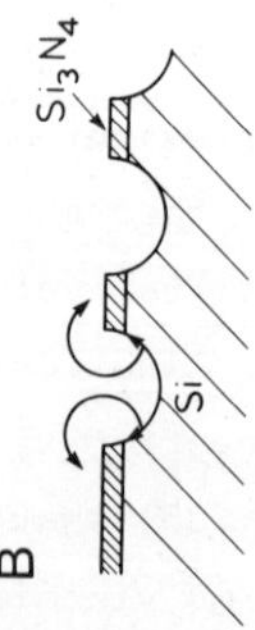
B
Si_3N_4
Si

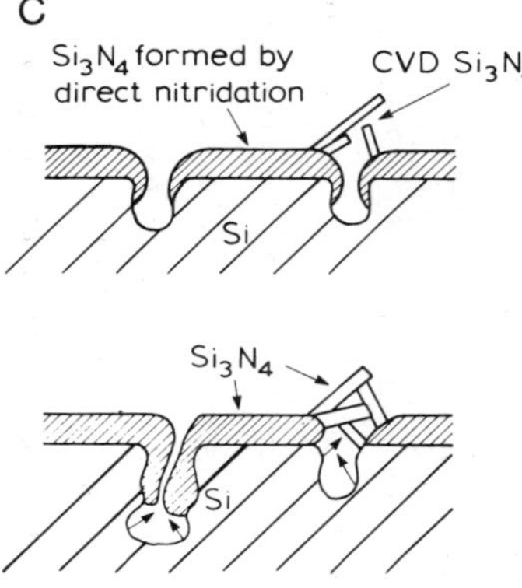

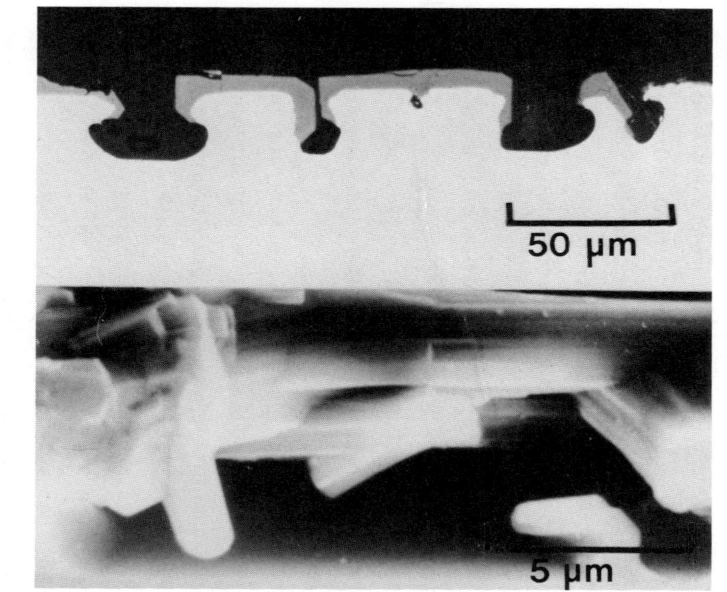

FIG. 6. Model for the nitridation of Si as proposed by Atkinson, Moulson, and Roberts (Ref. 40). A. Formation of Si_3N_4 nuclei on Si surface: areal density of nuclei increases as p_N increases and T decreases. B. Nuclei grow laterally and vertically: Si supplied to growth sites by combination of surface diffusion and evaporation. Rate of arrival of nitrogen to growth site determines reaction rate for $p_{N_2} \leq$ atm, at least. C. As free Si surface area decreases, surface diffusion distances increase and so direct nitridation slows. Reaction of Si in the vapour state to form a complex, which subsequently condenses, continues. 'Zero' reaction rate when reactants effectively separate.

have examined its effect in some detail. They argue convincingly that Fe contamination in the ppm range serves to disrupt by devitrification the silica film that is invariably present on Si, thereby exposing the Si surface and allowing removal of the oxide layer as SiO(g) via reaction (9). A further effect of Fe is to combine with Si to form Fe-Si liquids, some of which melt at temperatures as low as 1208°C; these liquids serve as media for the growth of β-Si_3N_4. The Fe-Si liquids also promote the growth of α-Si_3N_4, probably via the VLS mechanism proposed by Messier and Wong [43] and Gribkov et al. [51]. Figure 7, from Boyer and Moulson [63], illustrates and summarizes the processes discussed above.

The exact nature of the vapor species that participate in Si_3N_4-forming reactions has remained obscure. It has been pointed out by Blegen [35] and Lin [65], the latter on thermodynamic grounds, that direct nitridation of SiO(g) to form Si_3N_4 is impossible despite considerable speculation in the literature that such a reaction does occur. Moulson [40] addressed these issues, and, by thermodynamic calculations on reactions that were consistent with evidence in the literature, concluded that, depending upon conditions, both Si(g) and SiO(g) could participate in Si_3N_4-forming reactions. He pointed out that the vapor pressure of Si(g) is more than adequate to sustain nitridation rates observed under conditions under which that vapor species should predominate. The presence of H_2 in the nitriding gas has been observed to enhance the kinetics of nitriding of Si, and Moulson argues less convincingly than in the case above that the presence of H_2 allows the nitridation of SiO(g). The argument that he presents is based upon the thermodynamics of complex equilibria in the system Si-O-H-N. While it seems established unequivocally that one must consider the influence of small amounts of water vapor and oxygen in the nitriding system on nitridation kinetics, additional work appears necessary to clarify their roles in the process.

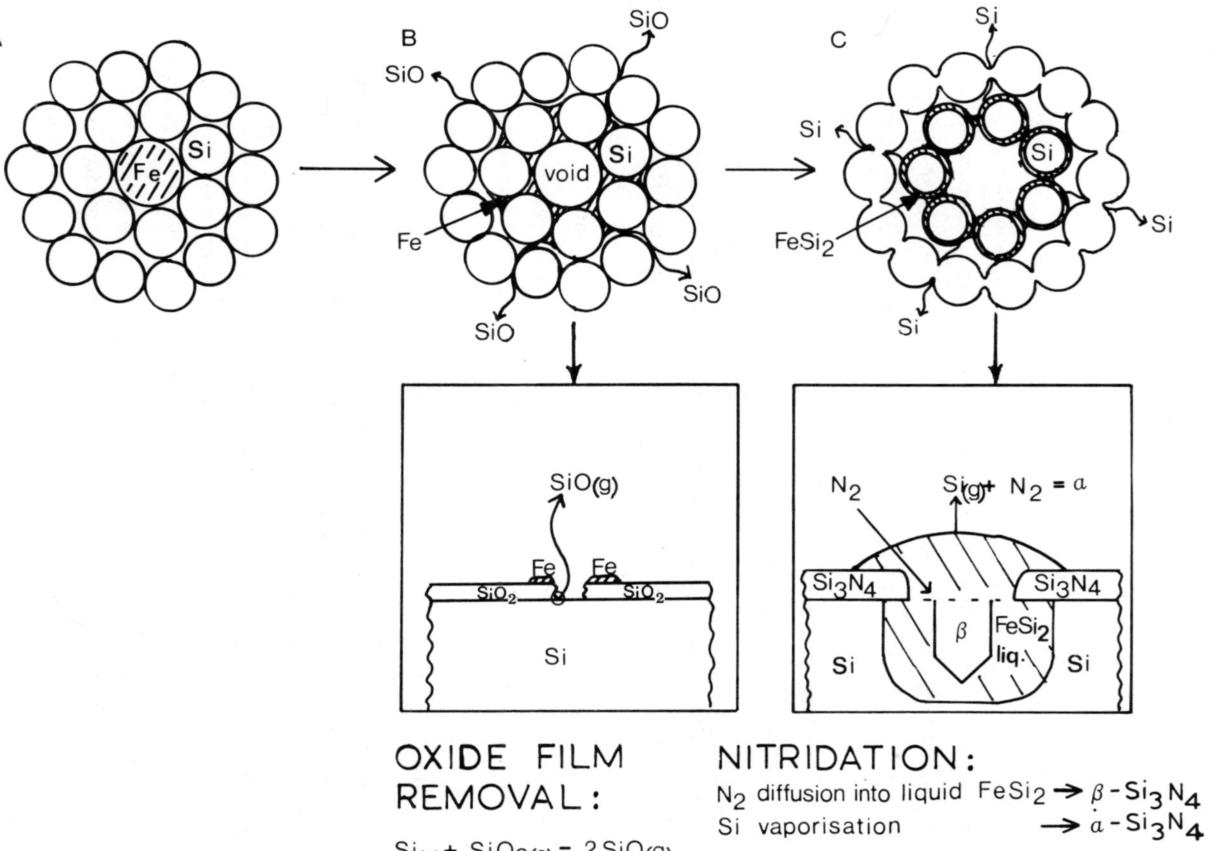

FIG. 7. Schematic representation of the role of iron in the nitridation of silicon. (From Ref. 63.)

Further studies of the kinetics of the nitridation of high-purity Si powder were reported by Inomata and Uemura [66] and Inomata [67]. The latter paper reports electron microscope studies of morphological features related to the kinetic observations in the former study.

The question of the exact conditions required for the formation of either α- or β-Si_3N_4, considered to some extent above, remains somewhat obscure. It has long been known that oxygen impurities in the nitriding atmosphere may be of some influence (e.g., Ref. 48), and the possible effects of other gaseous and solid impurities on favoring the formation of one phase or the other has been the subject of considerable interest [14,25,28-31,35,42-44,46,54,55,59,61,68-71]. It is apparent from the discussion in Sec. III, however, that α- and β-Si_3N_4 are very similar, and the fact that subtle differences influence the formation of one phase or the other is therefore not too surprising. Aside from the considerations already noted, it has been often observed that nitridation of Si at higher temperatures (∿1350°C or above) favors the formation of β-Si_3N_4, and that Si_3N_4 derived from vapor-phase reaction processes (powder or monolithic material produced by chemical vapor deposition) is invariably the α-form.

C. The α/β-Silicon Nitride Phase Transformation

As already mentioned in Secs. II and III, the exact relationship between the α and β forms of Si_3N_4 is still unclear. It has, however, long been established that heat treatment of α-Si_3N_4 at temperatures exceeding 1500°C converts it to β-Si_3N_4 [10,22,24]. In this section existing knowledge on the subject of the α/β transformation is reviewed with particular emphasis on the work of Messier et al. [23,72,73].

The importance of the α/β-Si_3N_4 phase transformation to the development of strength in hot-pressed Si_3N_4 was recognized by Coe et al. [74]. They reported that, although α-Si_3N_4 powder could be densified without transformation, high strength only obtained on material that had transformed. Lumby and Coe [75] further emphasized the importance of the transformation to hot-pressing, and it was suggested by others [76,77] that the strengthening may have been due to transformation-induced grain refinement. Until recently, however, little data have existed concerning the details of the transformation mechanism.

The role of oxide additives such as magnesia in the hot-pressing and phase transformation processes was first elucidated by Wild et al. [76]. They showed that magnesia reacted with silica (invariably present to some extent in silicon nitride) to form magnesium silicates that are molten at the temperatures (1600 to 1800°C) typically used in hot-pressing. They proposed that the transformation comprised decomposition of an oxynitride (α) to form the true nitride (β). As already mentioned in Sec. II, it now seems extremely unlikely that α-Si_3N_4 is an oxynitride, and the decomposition mechanism is therefore improbable.

It is only recently that quantitative data have been published on the kinetics of the α/β phase transformation, and, except for the work of Messier et al. [23,72,73], these data were obtained during experiments on the hot-pressing of Si_3N_4 powder [78-80]. Earlier, Coe et al. [74] found a linear relationship between percent β and log time for Si_3N_4 powder hot-pressed with 1 w/o magnesia at 1740°C, and several investigators gave a few results in similar experiments [31,68,81,82]

The most complete study of the kinetics of the α/β transformation was reported by Bowen et al. [80]. Their data were obtained in experiments of hot-pressing of commercial, high-α-Si_3N_4 powder containing 5 w/o magnesia at several temperatures between 1550 and 1750°C. They reported that the kinetics of the transformation were first-order, as defined by the equation:
temperatures between 1550 and 1750°C. They reported that the kinetics of the transformation were first-order, as defined by the equation:

$$\frac{d\alpha}{dt} = K\alpha \tag{10}$$

where K is the rate constant and α is the concentration of α-Si_3N_4 in the sample.

Drew and Lewis [83,84], from elegant electron microscopy studies of Si_3N_4 materials in various stages of hot-pressing, concluded that α/β-Si_3N_4 phase transformation occurs via a solid/liquid/solid mechanism, i.e., solution of α-Si_3N_4 in a liquid phase, and precipitation of β-Si_3N_4 from that liquid. This finding was corroborated by Messier et al. [23,72,73], Knoch and Ziegler [79], and Bowen et al. [80]. Also consistent with such a mechanism are evidence for the solubility of Si_3N_4 in molten magnesium silicates [85-87], and evidence that sintering in the Si_3N_4-magnesia system occurs via a liquid phase [80,86,88]. Greskovich and Prochazka [89] have presented evidence suggesting that the transformation can occur in high-purity powder in the absence of a liquid phase, contrary to the findings of Messier et al. [23,72,73] and Priest et al. [20] who found no evidence whatsoever for transformation of high-purity α-Si_3N_4 heated without additives to temperatures as high as 1800°C. This disagreement is difficult to rationalize, and it suggests that the transformation may be influenced by hitherto unrecognized experimental variables.

The appearance of commercial-purity silicon nitride powder (85% α-Si_3N_4) before and during the α/β transformation is illustrated in Figs. 8 and 9. The changes in morphology of the powder as the transformation progresses are very marked. The β-Si_3N_4 particles that result from the transformation have the typical hexagonal shape particularly evident in Fig. 9f.

It has been suggested from structural considerations that the α/β transformation should be reconstructive [24,90], and most of the available evidence indicates that it can be classified as a reconstructive transformation of secondary coordination [91]. Such transformations involve breaking bonds, but the first coordination is the same in each structure. Typically, such transformations are sluggish and only occur in the presence of solvents [92].

FIG. 8. Appearance of 99% pure, high-alpha (85%) silicon nitride powder. (From Ref. 73.)

It is still unclear whether α- and β-Si_3N_4 are true polymorphs and whether α-Si_3N_4 is metastable at all temperatures. The reverse transformation from β- to α-Si_3N_4 has never been observed experimentally, although Clancy [93] has presented some unconvincing circumstantial evidence suggesting that such a transformation occurred. All other evidence suggests that β-Si_3N_4 is the stable form above 1500°C or so; the proposed solution-precipitation mechanism of transformation is consistent with the expectation that a solution saturated with the unstable form (α) would be supersaturated with respect to the stable form (β) at a temperature above the transition point [92]. If α-Si_3N_4 is stable at some lower temperature, it may be that the kinetics are unfavorable for the reverse, β/α, phase transformation to occur. The resolution of some of these questions concerning the relationship between the two forms of silicon nitride appears to be an interesting area for further research.

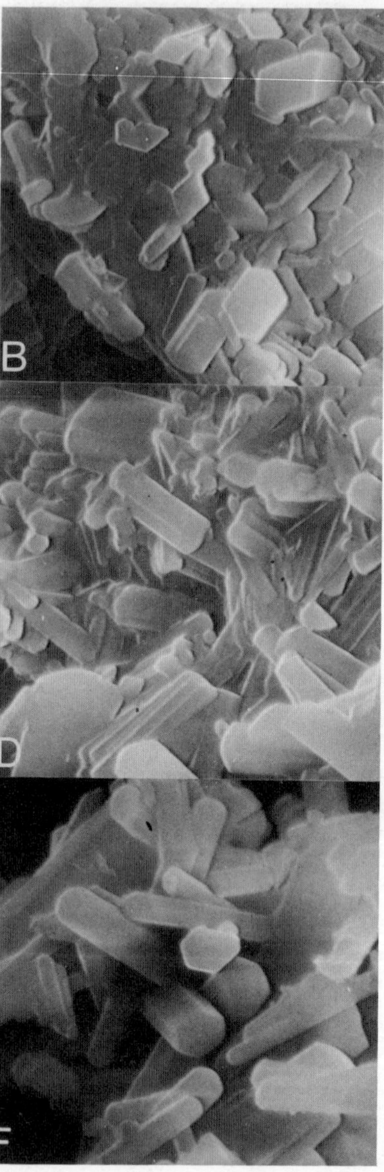

FIG. 9. Silicon nitride powder containing 10 w/o magnesia at various stages of transformation after heat treatment at 1600°C. (A) and (B): 0.5 hr, 46.8 w/o beta; (C) and (D): 1.0 hr, 64.1 w/o beta; (E) and (F): 4.0 hr, 100 w/o beta. (From Ref. 73.)

V. FABRICATION OF SILICON NITRIDE

A. Preparation of Silicon Nitride Powder

Although Si_3N_4 powder is readily available from a number of commercial vendors, relatively little literature exists on its preparation. For sintering and hot-pressing applications, it has been found particularly desirable to use powders comprising mostly α-Si_3N_4, and details of processes for manufacturing such powders have been regarded as proprietary by the vendors. The two methods most commonly used for the preparation of powders are direct nitridation of Si powder and vapor-phase processes such as the reaction between silicon tetrachloride and ammonia. Much of the literature already discussed in Sec. IV on Si-N_2 reactions is relevant to the synthesis of powder by the former route.

Work on the preparation of Si_3N_4 powder by nitridation of Si powder has been reported recently by Glaeser [94]. Emphasis in the paper is on the practical aspects of the process, i.e., producing the maximum amount of acceptable material in the minimum time.

Kato et al. [95] described the preparation of Si_3N_4 powder by vapor-phase reaction of silicon dichloride and ammonia. They examined the reaction products by several analytical techniques and noted that they were amorphous and slowly crystallized to α-Si_3N_4 with the evolution of excess nitrogen and hydrogen. Mazdiyasni and Cooke [96] and Mitomo et al. [97] reported work on powders made by the thermal decomposition of $Si(NH)_2$ The former authors also reported that the powder gradually transformed to α-Si_3N_4 upon thermal decomposition in vacuum at temperatures up to 1450°C. In addition, Mazdiyasni and Cooke gave results on materials hot-pressed from their powder. The $Si(NH)_2$ used by Mitomo et al. [97] was prepared by bubbling NH_3 into cooled $SiCl_4$.

Another route that may have commercial potential for the production of Si_3N_4 powder, namely synthesis from SiO_2, has been discussed by several authors [98-101]. This method, comprising nitridation of mixtures of SiO_2 and C, has been employed successfully to synthesize powders that are largely Si_3N_4. It is doubtful, however, that it will be feasible to produce high-purity Si_3N_4 powder in this manner.

B. Reaction-Bonded Silicon Nitride (RBSN)

1. *Introduction*

As mentioned in Sec. I, interest in silicon nitride was apparently renewed by the paper of Collins and Gerby [5] which emphasized the desirable physical properties of the material. That paper describes the formation of free-standing shapes from silicon powder by various means including dry-pressing, slip-casting, and extrusion and heating of those shapes at 1300°C in nitrogen to convert them to Si_3N_4. In this manner reaction-bonded silicon nitride (RBSN) was produced whose density was approximately 72% of theoretical. Collins and Gerby also reported various physical and chemical properties of their RBSN and included comparative data for other materials. Another early paper dealing with similar subject matter was published in Russia by Brokhin and Funke [102]. The paper of Popper and Ruddlesden [42], already mentioned in Sec. IV.A with regard to nitridation kinetics, also contains information that is still timely on forming methods and nitridation of shapes to produce RBSN.

Major contributions to the technology of RBSN have resulted from extensive work at the Admiralty Materials Laboratory (AML) in the United Kingdom initiated by Parr and his colleagues [103-107]. An early paper by Parr [103] described the preparation of RBSN bodies of density as high as 2700 kg m^{-3} and mentioned improved creep properties resulting from the inclusion of a fine dispersion of silicon carbide in the RBSN matrix. Other papers by Parr et al. [104,106, 107] reviewed the preparation and properties of RBSN, and one paper [105] described experiments intended to establish conditions which favored the formation of either α- or β-Si_3N_4.

Additional information on the preparation and properties of RBSN has been published by Rabeneau [108] and Godfrey [109]. Godfrey's paper stresses aerospace applications and discusses forming techniques, including pressing, flame-spraying, and plastic-binder dough forming, as well as the possibility of reinforcing RBSN with high-strength fibers. Additional information regarding these forming techniques and progress in the achievement of high-strength (300 MPa) RBSN appears in a paper by Brown et al. [110].

2. *Forming Methods*

The subject of methods of forming Si powder into shapes to be converted to RBSN has been addressed specifically in several papers. Brown [111] described in detail the preparation of Si bodies by flame-spraying. Valentine and Matkin [112] and Matkin et al. [113] discuss the fabrication of foamed silicon nitride and fabrication of RBSN by "ceramic/plastic technology." Slip-casting has been described by Harris [114] and Ezis [115] with the latter paper giving the most detailed description of the process yet published. Injection-molding, a forming method that has outstanding potential for high-volume production, has been the subject of two recent papers [116,117]. An automatic control system for that process, which comprises injection under pressure of an Si-thermoplastic mixture into a heated metal die, was described by Johnson and Mohr [116].

3. *The Reaction-Bonding Process*

The effects of processing conditions on the evolution of microstructure and on the properties of RBSN have been extensively investigated. Although there is still some uncertainty as to the details of the mechanisms of the process, considerable progress has been made in producing high-strength (up to 400 MPa) material. Some information not cited in the section on the nitridation of silicon but pertinent to the reaction-bonding process appears in publications by Atkinson and Moulson [118] and Lin [119,120]. The former paper concerns the effects of iron additions on the nitridation of Si and their effect on the microstructure of the resulting RBSN. The papers by Lin [119, 120] report the effects of various metallic additives on nitridation. An article by Amato et al. [121] discusses the influence of raw materials and processing variables on the preparation of RBSN.

Dalgleish and Pratt [122] studied the development of microstructure during nitridation at temperatures ranging from 1250 to 1500°C by optical and electron microstopy. While the maximum nitridation temperature that was used is considerably higher than is normal fabrication practice, the microstructures shown reflect reasonably well what is observed in RBSN materials. At temperatures below the melting point of Si, the pores of the Si compact become filled with needles which eventually form a matte of α-Si_3N_4. Above the melting point of Si, the matte becomes denser and crystallites of β-Si_3N_4 grow into the molten Si. The resultant structure comprises discrete grains of β-Si_3N_4 enmeshed in a matte of α-Si_3N_4.

The most comprehensive investigation by far on the nitridation process and its effects on properties is described in a series of papers by Jones, Lindley, and coworkers [123-133] from AML.

In the first four papers [123-126] of this series, Jones and Lindley discuss in detail relationships that they found among strength, density, and nitrogen weight gain for Si powder compacts at various stages of nitridation. Among their conclusions were, as expected, that strength was proportional to nitrogen weight gain and to the green density of the Si compact. A remarkable conclusion of this work, however, was that the strength of a partially nitrided compact depended only upon bulk density, and not upon the relative amounts of the phases (Si and Si_3N_4) in the compact. Jones and Lindley point out that this strength-density relationship can be used in a number of ways to analyze and optimize the reaction-bonding process. For example, it can be used for comparison of the nitridation behavior of different starting Si powders and for prediction of the maximum nitrided strength obtainable with each. The relationship is also useful in assessing the effect of other variables such as compaction conditions and prenitriding heat treatments on the process. The papers by Jones and Lindley [123-126] contain considerable experimental data in support of the conclusions outlined above.

Additional papers [127-130] by Elias, Jones, and Lindley consider the effects of hydrogen and oxygen contamination on the nitriding process as well as effects resulting from the use of a "static" or "flowing" atmosphere. Elias et al. [127] concluded that α- and β-Si_3N_4 form via reactions between N_2 and, respectively, SiO and Si, with the source of SiO being the SiO_2 contamination invariably present in Si powder. They also point out that reaction rates, phase composition, and strength are all dependent upon the properties of the starting Si powder and upon whether the nitriding atmosphere is "flowing" or "static." The latter point is subtle but critical; much of the variability in earlier results could have stemmed from such effects.

Elias and Lindley [128] demonstrated that, contrary to previous thought, the presence of oxygen and water vapor contamination in relatively high amounts in the nitriding atmosphere did not significantly degrade the strength of RBSN. They also found some interesting but puzzling correlations between the content of water vapor in the nitriding gas and variability of strength as well as "apparent" crystallite size. In addition, they concluded that the composition of α-Si_3N_4 is fixed and that α-Si_3N_4 contains little, if any oxygen.

The effects of nitridation under "flow" or "static" conditions on the strength of RBSN were discussed specifically in two papers by Jones and Lindley [129,130]. They showed that material produced under "static" nitrogen was stronger than that prepared in "flowing" nitrogen and that the addition of hydrogen to the atmosphere in a flowing system eliminated the strength difference.

Further publications by Jones et al. [131-133] report results of mechanical property and fracture mechanics measurements and interpret these results in terms of the various nitridation phenomena previously discussed. It is shown that elastic modulus, modulus of rigidity, and strength vary linearly with nitrided density, and also that, after an initial period, critical defect size decreases continuously with increasing nitrided density. It is further proposed that, at any particular density, the critical defects are larger under "flow" than under "static" conditions. Figure 10 summarizes the work and shows schematically the model of the development of structure in RBSN as envisaged by Jones et al. [131].

4. *Fabrication-Property Relationships*

In addition to the AML work already discussed, a number of other investigations have been concerned with the relationships between fabrication conditions and the resulting properties of RBSN.

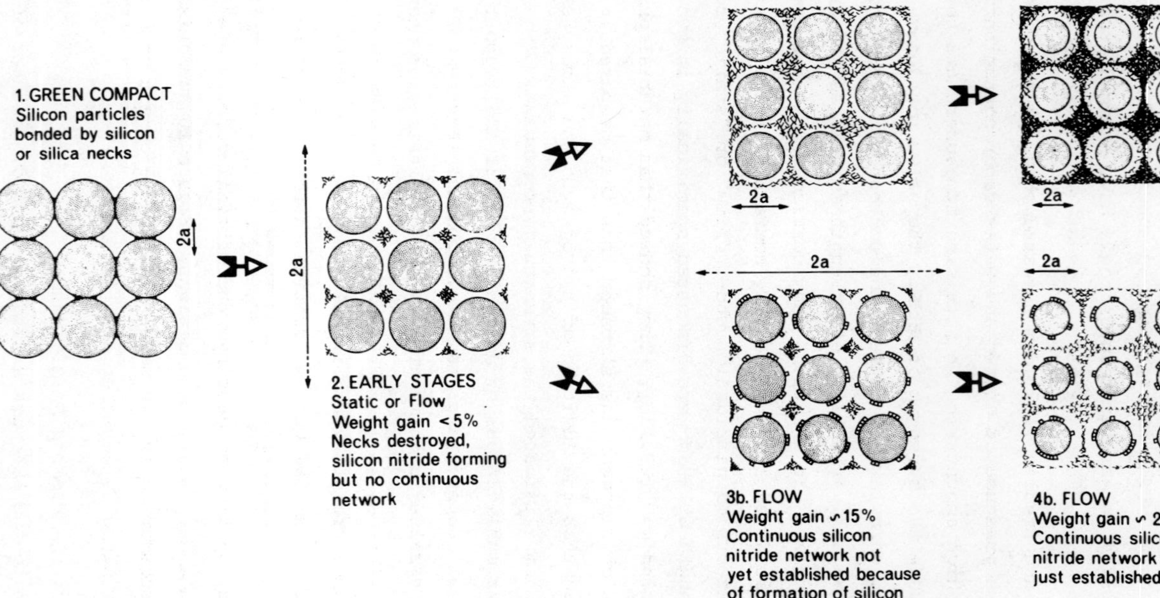

FIG. 10. Schematic model of the development of the structure of reaction-bonded silicon nitride. (From Ref. 131.)

A significant contribution on this subject was made by Evans and Davidge [134] who proposed that the strength-limiting defects in RBSN were large pores resulting from the melting of unreacted Si in the material during the final stage of nitridation. This particular type of defect, while relatively unimportant in nitriding small specimens of the type used by Jones, Lindley, and coworkers [123-133], becomes extremely important when nitriding large amounts (several kg) of material where control of the exothermic reaction is crucial to successful processing. Godfrey and Lindley [135], in a study of the effects of processing on the strength of RBSN, were unable to confirm the conclusion of Evans and Davidge [134]. A study by Messier and Wong [136], however, was in substantial agreement with the latter. In addition, the work of Messier and Wong indicated that the use of fine Si powder and low maximum nitriding temperature increased strength. They further demonstrated that material prepared from high-purity (99.99%) Si was at least as strong as that fabricated from impure Si powder.

Godfrey [137] reported on the effects of Fe and Al impurities on the formation of RBSN. He found that additions of Al_2O_3 and SiC reduced strength and that the former addition increased the content of β-Si_3N_4. Godfrey and Pitman [138] found that the strength of RBSN was independent of maximum nitriding temperature (1350 vs. 1450°C), but was dependent upon porosity.

The effects of nitriding atmosphere and nitriding cycle on the strength of RBSN were investigated by Mangels [139,140]. He found [139] that addition of hydrogen to the nitriding atmosphere altered the microstructure and improved properties at both room and high temperatures. He also investigated extensively the effect of the time-temperature cycle in nitriding on the properties of the resulting material on quantities similar to small production lots and concluded that control of the exothermic reaction was critical for control of the process [140]. A recent paper on the effect of Si particle size on the microstructure and mechanical properties of RBSN has been published by Heinrich [141].

5. Nitridation Control Systems

Systems for the controlled nitridation of large batches of Si powder shapes have been described by Wong and Messier [142] and Mangels [143]. Both systems are similar and are readily adaptable to automation of the process. The systems are based upon detection of the beginning of the consumption of nitrogen by the mass being nitrided and using a signal from the detector (i.e., a pressure or flow transducer) to control the rate of temperature increase in the nitriding furnace. By this means the exothermic reaction is controlled and a uniform, reproducible material results.

6. Infiltration of RBSN

A few attempts have been made to improve the properties of RBSN, which typically has considerable open porosity, by coating or infiltration. Inomata [144] described experiments in which Si layers 40-100 μm thick were formed by heating Si-coated Si_3N_4 in nitrogen at 1430-1450°C. This procedure greatly improved oxidation resistance. Leimer and Gugel [145] reported infiltration experiments in which various alloys were used. More recently, Mazdiyasni et al. [146] gave results on the infiltration of RBSN with organosilicon liquids that were subsequently decomposed to produce additional SiC or Si_3N_4 in the pores.

C. Hot-Pressed Silicon Nitride (HPSN)

Hot-pressed silicon nitride is the strongest form of that material and probably the highest-strength ceramic material of any kind available in quantity. The hot-pressing process comprises heating under pressure a mixture of Si_3N_4 powder (preferably high in α phase content) and a sintering aid. Fabrication is done by loading the powder into a graphite die and hot-pressing at pressures of around 30 MPa for a few hours at temperatures in the range 1600 to 1800°C. The first report of the preparation of Si_3N_4 by this route was that of Deely et al. [147]; they successfully used MgO and Mg_3N_2 as sintering aids. Aspects of the fabrication of hot-pressed silicon nitride have been reviewed in several publications [148-151]. Two of these have been concerned with HPSN for turbine applications; Gugel et al. [149] discussed German work in this area, and Bratton et al. [151] summarized the extensive effort by Westinghouse and Norton Company in the United States on the development of HPSN vanes.

Following the work of Deely et al. [147] a substantial effort ensued on hot-pressing in the system Si_3N_4-MgO. This system has been important for two reasons: (1) it has served as a model system for elucidating mechanisms of sintering with MgO and other additives; and (2) it has been the system most often used commercially for the fabrication of HPSN. Coe et al. [74] and Lumby and Coe [75] described in some detail their work on the fabrication and properties of HPSN prepared with MgO additions, and further studies were reported by Weston and Carruthers [81] and Oyama and Kamigaito [152]. The first investigations that elaborated on the mechanisms of the process were those of Wild et al. [76] and Colquhoun et al. [153]. They explained that the MgO additive combined with the SiO_2 invariably present as a contaminant in Si_3N_4 powder to form magnesium silicate liquid phases that serve as sintering media. They further point out that, when the piece is cooled after hot-pressing, the liquid becomes a glass; the resulting HPSN structure consists of grains of β-Si_3N_4 bonded together by a thin glassy layer at the grain boundaries. As noted by Wild et al. [76], this grain boundary phase accounts for the relative lack of high-temperature (>1200°C) strength of this material.

Hot-pressing reactions and mechanisms in the system Si_3N_4-MgO have been addressed further in several papers. Drew and Lewis [83] presented results of an elegant electron microscopy study of the microstructures of Si_3N_4 compacts at various stages of hot-pressing. This work emphasized in particular the role of the α/β-Si_3N_4 phase transformation in the process and it gave persuasive evidence of the importance of solution-precipitation mechanisms. Terwilliger and Lange [154] further defined the role of MgO in liquid-phase formation and also pointed out that the phase transformation was unnecessary for densification. Examinations of the microstructure of HPSN were also reported by Nuttall and Thompson [85] and by Osipova and Pogorelova [155].

The most detailed analysis to date of mechanisms of densification in the hot pressing of Si_3N_4 with MgO appears in a series of papers from the University of Leeds, U.K., by Brook et al. [78] and Bowen et al. [80,156]. They propose a two-stage process: the initial stage involves particle rearrangement, liquid-enhanced above 1550°C; and the second and final stage follows the Coble hot-pressing equation with the rate of densification proportional to the amount of additive.

The relationship between mechanical properties and processing has been the subject of a number of investigations. Iskoe et al. [157] reported that, among a number of impurities investigated, CaO and alkaline oxides were primarily responsible for high-temperature strength degradation in HPSN fabricated with MgO. Knoch and Ziegler [79,158] have investigated the effects of MgO content and a number of other processing variables on the structure and strength of HPSN. Other investigations of the correlation between processing and properties of this material have been reported by Uy et al. [159] and by Bowen and Carruthers [160].

Recognizing the need for a hot-pressing additive that would give high temperature mechanical properties superior to those of HPSN formed with MgO, Gazza [161,162] discovered that Y_2O_3 would perform such a function. His work stimulated further research on the use of that additive as well as on identification of oxide and nitride compounds in the system Si-Y-O-N. Subsequent papers on mechanisms of hot-pressing and properties of HPSN fabricated with Y_2O_3 were published by Tsuge et al. [163], Weaver and Lucek [164], and Bowen et al. [165]. Tsuge and Nishida [166] discussed fabrication of HPSN with concurrent additions of Al_2O_3 and Y_2O_3. Compounds and phase relationships in the system Si_3N_4-Y_2O_3 have been discussed by Wills [167], Tsuge et al. [168], and Rae et al. [169]. A recent article by Gazza et al. [170] has resolved satisfactorily questions that had been raised about the thermal instability under certain conditions of HPSN containing Y_2O_3. It should be noted that research is still very active with respect to the use of the Y_2O_3 additive, and perusal of the recent literature would be essential to obtaining a complete assessment of current understanding of this subject.

Some effort has been expended on the use of hot-pressing aids other than MgO and Y_2O_3. Rice et al. [171] and Rice and McDonough [172] reported that Zr compounds including ZrO_2, ZrN, ZrC, and zircon were effective sintering aids, and Huseby and Petzow [173] reported success using CeO_2, Ce_2O_3, and La_2O_3. Recent papers by Weston et al. [174] and Weston and Pratt [175] discussed the preparation and properties of HPSN with multicomponent oxide additions that formed glassy phases that were subsequently recrystallized by annealing.

D. Sintered Silicon Nitride

While hot-pressed silicon nitride has desirable physical and chemical properties for demanding applications, complex shapes are difficult to make by hot-pressing, and the process is expensive. For those reasons, there has developed considerable interest in fabricating silicon nitride by the conventional ceramic fabrication technique of sintering without applied pressure. The same oxide sintering aids mentioned in the previous section are being investigated for the preparation of Si_3N_4 bodies by pressureless sintering. A major problem in the fabrication of Si_3N_4 by this route is that the sintering temperatures required, $\sim$1700-1800°C, are in the range where thermal decomposition becomes a problem. In order to prevent decomposition, one must either use an apparatus in which a nitrogen overpressure can be maintained or carefully select a combination of sintering conditions and additive by which the processing can be carried out in nitrogen at atmospheric pressure.

The first reported attempt to fabricate Si_3N_4 by pressureless sintering is that of Terwilliger [176]. He indicated that Si_3N_4 powder compacts containing 5 w/o MgO could be sintered to 90% of theoretical density at 1650°C. Other conditions were unspecified. As indicated by the date (1974) of Terwilliger's work, this is a relatively new research area. The only review on the fabrication of Si_3N_4 by pressureless sintering had been presented by Gazza [177]. He provided a brief survey of the state-of-the-art in this rapidly developing area of technology.

Additional results on the sintering of Si_3N_4 powder with MgO additions were given by Terwilliger and Lange [88]. They were able to prepare bodies that were 85-90% of theoretical density by sintering in nitrogen at atmospheric pressure at temperatures from 1500 to 1750°C. The products that they obtained, however, were of relatively low strength. Improved density and strength were reported for Si_3N_4-MgO materials fabricated at higher temperatures and N_2 pressures [178-180]. Sleptsov et al. [178] gave results on sintering at N_2 pressures from 1.0 to 3.0 MPa, and Mitomo et al. [179,180] prepared material of 95% of theoretical density by sintering at 1800°C in N_2 at 1.0 MPa. As already indicated for hot-pressing, the general consensus [88,178-180] is that pressureless sintering occurs via a liquid-phase mechanism.

More recent work on sintering has emphasized the use of additives other than MgO. The general objective of such work is to find additives that are effective as sintering aids but that also produce material with optimum room- and high-temperature properties; mechanisms of high-temperature strength degradation are usually similar to those of sintering, so the choice of additive is often a compromise. Priest et al. [181] investigated the sintering in N_2 at 2.1 MPa and at temperatures from 1800 to 1950°C of Si_3N_4 with respective additions of CeO_2, Y_2O_3, and MgO and obtained some bodies that were highly dense. A recent publication by Mah et al. [182] discusses work on compounds in the system Ce-Si-O-N and considers the role of Ce orthosilicate in the densification of Si_3N_4. Greskovich et al. [183] reported on the sintering of relatively pure Si_3N_4 with and without an unspecified "nonoxide" additive and concluded that the additive was essential to densification. A recent publication by Greskovich and O'Clair [184] considers sintering with MgO and Y_2O_3 additives with emphasis on effects of impurities.

The use of multicomponent oxide additives has been investigated by several workers. Masaki and Kamigaito [185] found that spinel ($MgAl_2O_4$) was more effective than either MgO or Al_2O_3 used individually. Mitomo [186] employed mixtures of Al_2O_3 and Y_2O_3, and Oda et al. [187] used combinations of BeO-MgO-CeO_2.

As already indicated, considerable research is currently in progress on the fabrication of Si_3N_4 by pressureless sintering. Evidence of this interest is reflected in programs of recent conferences, and it is to be expected that more data on this subject will soon appear in the open literature.

E. Chemical Vapor Deposition

Chemical vapor deposition (CVD) techniques have potential for the preparation of fully dense, high-purity Si_3N_4 at relatively low temperatures. A book by Milek [8] reviews techniques widely employed by the electronics industry for producing CVD Si_3N_4 films for applications such as insulating films and diffusion masks. The same book also discusses film preparation by sputtering, glow discharge reactions, and direct evaporation. These techniques, however, have proven to be less successful than CVD for thin film production.

The most commonly used techniques for CVD of thin films involve chemical reactions between silane or silicon halides and ammonia. The addition of H_2 is sometimes, but not always, beneficial. Typically, the reactions are carried out at temperatures from 800 to 1100°C. The insulating films that are prepared are generally amorphous and result from the following reactions:

$$3SiH_4(g)\ \text{(silane)} + 4NH_3(g) = Si_3N_4(s) + 12H_2(g) \tag{11}$$

$$3SiCl_4(g) + 4NH_3(g) = Si_3N_4(s) + 12HCl(g) \tag{12}$$

$$3SiF_4(g) + 6NH_3(g) = Si_3N_4(s) + 2NH_4F(g) + 10\ HF(g) \tag{13}$$

$$3SiH_4(g) + 3N_2H_4(g)\ \text{(hydrazine)} = Si_3N_4(s) + 2NH_3(g) + 9H_2(g) \tag{14}$$

$$3SiHCl_3(g)\ \text{(trichlorosilane)} + 4NH_3(g) = Si_3N_4(s) + 9HCl(g) + 3H_2(g) \tag{15}$$

Properties of deposited films such as refractive index, density, chemical composition, and etching rate may vary considerably with variations in deposition parameters, e.g., time, temperature, and flowrate.

Attempts to prepare thick, monolithic deposits of Si_3N_4 by CVD have been reported by Galasso et al. [26,188]. They employed the reaction between SiF_4 and NH_3 in the temperature range 1100 to 1550°C to deposit plates and cylinders 1.0-1.5 mm thick on carbon substrates. The substrates were subsequently removed by oxidation, and free-standing Si_3N_4 shapes were thus produced. Also produced were crucibles with walls 3 mm thick. The resulting materials were found by x-ray diffraction to be either amorphous or crystalline. In the latter case they consisted of α-Si_3N_4. The CVD materials were found to be extremely resistant to oxidation as well as to chemical attack by KOH. Galasso et al. [26,188] also deposited CVD coatings on other substrates including Al_2O_3, BN, Mo, SiO_2 in various forms, and on reaction-bonded and hot-pressed Si_3N_4.

Kijima et al. [189] investigated the growth of Si_3N_4 from gaseous mixtures of $SiCl_4$, H_2, and N_2 in the temperature range 1000 to 1800°C. This particular system was chosen because of the lack of chemical reactions at room temperature; mixing can therefore be done before the gas enters the reaction vessel. The products that were obtained varied considerably with the reaction temperature. Between 1100 and 1200°C, white, feltlike masses of fibers formed. These were amorphous, but were identified as Si_3N_4 by infrared absorption spectroscopy. In the temperature range 1300-1400°C, yellow polycrystalline masses comprising a mixture of α- and β-Si_3N_4 were obtained; the α-Si_3N_4 crystals showed preferred orientation with their (002) directions being perpendicular to the substrate. At higher temperatures (1400-1500°C), whiskers of α- and β-Si_3N_4 grew simultaneously, and from 1500 to 1600°C prismatic single crystals (0.2-0.5 mm long) of α-Si_3N_4 resulted. These crystals were transparent, colorless to brownish red, and again oriented with the (002) direction perpendicular to the substrate. Kijima et al. also investigated the effect of N_2/H_2 ratio on growth and concluded that, at 1600°C, varying the N_2 partial pressure from 30 to 80 kPa had little effect on the morphology of the resulting Si_3N_4. When the N_2 partial pressure was lowered to 2.2 kPa at the same temperature, a small amount of SiC formed. It was concluded from the investigation of Kijima et al. that the net chemical reaction for the formation of Si_3N_4 at 1500°C is

$$3SiCl_4(g) + 6H_2(g) + 2N_2(g) = Si_3N_4(s) + 12HCl(g) \quad (16)$$

Recent work on the CVD process is described in a series of papers by Niihara and Hirai [190-193]. Those investigators deposited Si_3N_4 on graphite substrates from gaseous mixtures of $SiCl_4$, NH_3, and H_2. The initial reaction product, $Si(NH)_2$, forms as follows:

$$SiCl_4(g) + 6NH_3(g) = Si(NH)_2(s) + 4NH_4Cl(s) \quad (17)$$

During subsequent heat treatment, the $Si(NH)_2$ polymerizes and eventually decomposes to form α-Si_3N_4 via the following sequence of reactions:

$$6[Si(NH)_2]_n \xrightarrow[-2NH_3]{400^\circ C} 2[Si_3(NH)_3N_2]_n \xrightarrow[-NH_3]{650^\circ C} 3[Si_2(NH)_2]_n \xrightarrow[-NH_3]{1200^\circ C} 2\alpha\text{-}Si_3N_4 \quad (18)$$

These reactions occur at temperatures lower than the ones used for deposition; they are listed because they represent intermediate steps that occur at high reactant concentrations. Under such conditions, Si_3N_4 powder forms in the gas phase before deposition. A consequence of this is that deposition rates actually become lower when the pressure of the reactant gases becomes excessively high.

Niihara and Hirai were able to produce relatively thick (4.6 mm) deposits at substantial growth rates (0.73 mm hr^{-1}). The material was amorphous under all conditions when deposited at 1100 and 1200°C. At 1300°C, amorphous material resulted when the gas pressure in the system was in the ranges 1.3-5.3 kPa and 10.5-39.5 kPa. At the same temperature, however, crystalline α-Si_3N_4 was produced at pressures from 5.3 to 9.2 kPa.

The CVD materials produced by the methods discussed thus far have a number of common characteristics. The crystalline α-Si_3N_4 deposits are of densities (3.15-3.18 g cm^{-3}) near to or at the theoretical value of 3.18 g cm^{-3}. In addition, the oxygen content of the crystalline material is low, typically 0.3 w/o or less. The densities of the amorphous deposits depend strongly upon deposition conditions, and values as low as 2.60 g cm^{-3} have been obtained for deposits formed at 1200°C and 5.3 kPa gas pressure [191].

Work has also been reported on the CVD of Si_3N_4 from reactions involving SiO(g), the SiO being generated by heating mixtures of SiO_2 and C. Knippenberg and Verspui [194] prepared deposits comprising a mixture of Si_3N_4 and SiC from reactions in the system SiO_2-C-H_2-N_2 at 1400°C. Kijima et al. [195] investigated the preparation of CVD Si_3N_4 by a technique in which N_2 gas was passed over a graphite crucible containing a mixture of SiO_2 and C and deposition occurred on a Si substrate. Films of deposit began to form at 1200°C and α-Si_3N_4 was obtained at 1400°C. They also reported that, above 1500°C, SiC formed on the graphite crucible.

VI. MECHANICAL PROPERTIES

The considerable interest in silicon nitride as a high-temperature structural material stems from its unique combination of properties. In addition to having outstanding resistance to corrosion and thermal shock, silicon nitride is one of the strongest ceramic materials currently available. Further desirable characteristics are its excellent high-temperature strength and creep resistance.

To give a comprehensive review of the mechanical properties of silicon nitride would be a formidable task beyond the scope of the present chapter which emphasizes the chemistry of the material. This section will therefore be limited to a survey of a few relevant publications on the subject and to a brief summary of representative properties of Si_3N_4 materials.

An excellent, comprehensive review of the mechanical properties of various Si_3N_4 materials was presented by Edington et al. [196,197]. Another review emphasizing HPSN has been published by Lange [198]. The only deficiencies in these reviews are those shared by all such efforts, including the present one; in a dynamic field of research and development, new data are being generated continuously. One must therefore consult the subsequent literature for current property data.

Literature on the mechanical properties of RBSN includes a paper by Evans and Davidge [134] which appears to be the first to discuss in detail relationships between structure and mechanical properties of that material. Also discussed is the effect of oxidation on strength, a subject that is of considerable current interest. The high-temperature mechanical properties of RBSN have been discussed in several papers [199-202]. Mangels [199] considered the effects of impurities and nitriding conditions on creep rates and concluded that creep was reduced by reducing the Ca impurity level and by nitriding in an atmosphere containing H_2. Creep results were also given by Washburn and Baumgartner [200] and by Thuemmler et al. [201]. The latter concluded that creep deformation was accelerated in air as compared to vacuum because of the formation of oxide phases in the former environment. Further data on hot strength, creep, and stress rupture were given by Mangels [202]. He reported no slow crack growth in RBSN at temperatures up to 1400°C.

The mechanical properties of HPSN have been discussed in numerous publications. Of particular concern has been the high-temperature (>1200°C) strength of that material. As already mentioned, any additive that is effective as a sintering aid tends to degrade high-temperature properties; a considerable effort has therefore been conducted on achieving the optimum compromise between the two. Gazza [161] was the first to report that the use of Y_2O_3 rather than MgO as a sintering aid improved the high-temperature strength of HPSN. Data on fatigue, creep, and stress rupture of HPSN containing MgO were reported by Kossowsky [203] and Kossowsky et al. [204]. They concluded that high-temperature strength was controlled by grain boundary sliding and they developed a model for the deformation process. Additional data on the mechanical properties of HPSN containing Y_2O_3 were given by Gazza [162] and by Tsuge et al. [205]. The latter showed that crystallization of the grain boundary phase increased high-temperature strength. A recent paper by Weaver and Lucek [164] discusses optimization of HPSN materials containing Y_2O_3 with emphasis on high-temperature mechanical properties and oxidation resistance.

It has become increasingly apparent that the undesirable oxidation and high-temperature strength behavior sometimes observed in HPSN are associated with the amount of sintering aid (e.g., MgO or Y_2O_3) used in the hot-pressing process. Lange [206] argues persuasively that the mechanical and oxidation characteristics of these materials are governed by phase relationships in the systems Si_3N_4-SiO_2-oxide additive. Desirable properties can therefore be achieved by maintaining the composition of the system in the region of the phase field which precludes the formation of unwanted compounds.

Table 5 lists some representative data on room-temperature mechanical properties of several silicon nitride materials. The data for RBSN and HPSN (fabricated with Y_2O_3) were taken from a report by Quinn [207]. The data for HPSN fabricated with MgO are from the summary report by Miller et al. [208] which gives the results of an extensive program directed toward the use of that material in a large gas turbine. The data for sintered Si_3N_4 fabricated with Y_2O_3 were obtained from a recent report by Smith and Quackenbush [209]. It should be emphasized that the selection of values to include in a compilation such as is given in Table 5 is always somewhat arbitrary; most of the materials are still under development, and the values

TABLE 5

Room-Temperature Mechanical Properties of Si_3N_4 Materials

	Unit	RBSN [207]	HPSN (MgO) [208]	HPSN (Y_2O_3) [207]	Sintered Si_3N_4 (Y_2O_3) [209]
Density	kg m^{-3}	2770	3200	3370	3240
Young's Modulus, E	GPa m^{-2}	210	300	310	275
Poisson's Ratio, μ	--	0.22	0.25	0.27	0.23
Shear Modulus, G	GPa m^{-2}	86	120	122	112
4-Point bend strength	MPa	288	760	920	665

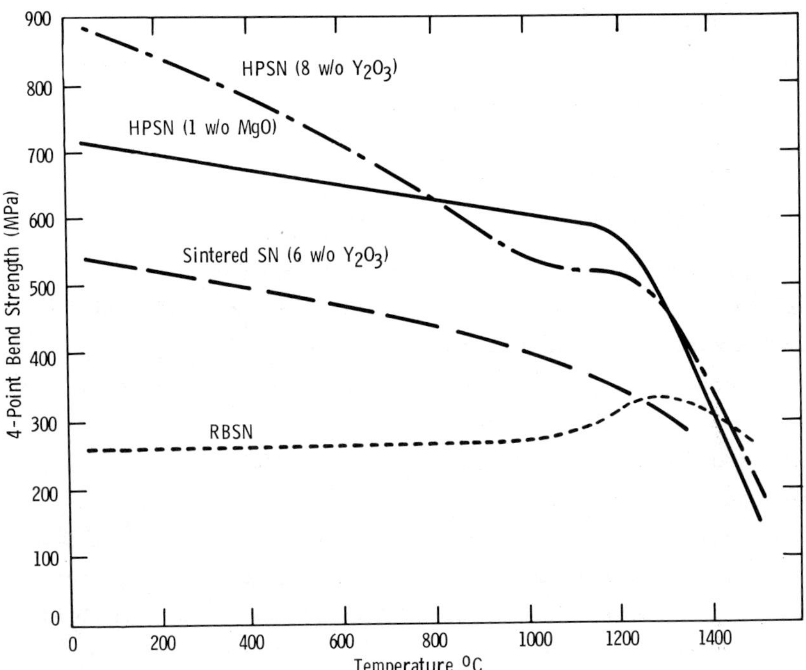

FIG. 11. Four-point bend strength vs. temperature for various silicon nitride materials. (From Ref. 210.)

obtained for flexural strength depend somewhat upon the size of the specimen used and on the details of the testing procedure. It is felt, however, that the data in Table 5 represent reasonably well the state-of-the-art at present.

Figure 11 shows the temperature dependence of the flexural strength of four Si_3N_4 materials. The data in Fig. 11 were taken from a report by Larsen and Walther [210] who are conducting a program on the compilation of test data on a variety of materials of interest for turbine applications. For reasons already given, the room-temperature strength values in Fig. 11 differ slightly from those listed in Table 5. Since the results in Fig. 11, however, were obtained from tests in the same apparatus, they provide a valid comparison among the several materials. The behavior shown in Fig. 11 is typical; the materials which were fabricated with sintering aids show a dropoff in strength with increasing temperature, whereas the strength of RBSN is relatively low, but constant or slightly increasing with increasing temperature.

VII. THERMAL PROPERTIES

A. Thermal Expansion

Interest in silicon nitride was stimulated some years ago by the paper of Collins and Gerby [5] which noted the low coefficient of thermal expansion of that material. That particular property is of great importance technologically, for the combination of low thermal expansion and high strength makes silicon nitride one of the most thermal-shock-resistant ceramic materials available for demanding high-temperature applications.

A detailed analysis of the thermal expansion behavior of silicon nitride has been provided by Henderson and Taylor [90]. In that paper, previous work is reviewed and thermal expansion data obtained by x-ray diffraction methods for α- and $\beta\text{-Si}_3N_4$ are presented. According to Henderson and Taylor, the thermal expansion from 20 to 1000°C of the two forms of silicon nitride can be represented by equations of the form

$$y = y_0(1 + bT + cT^2) \tag{19}$$

where y is the cell parameter, y_0 the cell parameter at 0°C, T the temperature (°C), and b and c regression coefficients. For $\alpha\text{-Si}_3N_4$, a direction:

$$y = 7.4467 + 1.836 \times 10^{-6}T + 1.774 \times 10^{-9}T^2 \tag{20}$$

and for c direction:

$$y = 5.6158 + 2.589 \times 10^{-6}T + 1.111 \times 10^{-9}T^2 \tag{21}$$

For $\beta\text{-Si}_3N_4$, a direction:

$$y = 7.5999 + 0.795 \times 10^{-6}T + 2.432 \times 10^{-9}T^2 \tag{22}$$

and for c direction:

$$y = 2.9059 + 1.955 \times 10^{-6}T + 1.762 \times 10^{-9}T^2 \quad (23)$$

Henderson and Taylor interpret the thermal expansion behavior of Si_3N_4 in terms of its crystal structures, i.e., α-Si_3N_4 having a moderately strained structure and β-Si_3N_4 only a slightly strained one. The anisotropy of thermal expansion is attributed to the irregularity of the tetrahedra forming the structure, and it is suggested that it results from either or both of two causes: (1) a change in the relative distortion in the a and c directions as temperature increases; and (2) the lower expansion coefficients of bonds parallel to the (001) plane as compared to those of bonds at angles to that plane.

The literature contains numerous data on dilatometer measurements of thermal expansion coefficients of various Si_3N_4 materials. These measurements in general do not take into account anisotropy and they were made on materials of various purity with or without sintering aids, unreacted silicon, etc. It is therefore not surprising that a range of values exists, and it is recommended that the literature be consulted when a precise value is required for a particular material. For general purposes, it appears reasonable to accept the mean expansion coefficient of $3.0 \times 10^{-6}\ °C^{-1}$ for the temperature range 20-1000°C as suggested by Moulson [40].

B. Specific Heat

The results of measurements of specific heat values for several Si_3N_4 materials appear in Fig. 12. Included in the figure are data for hot-pressed [208,211], sintered [209], and reaction-bonded [210] Si_3N_4. The results reported by George [211] were obtained on one hot-pressed and two reaction-bonded materials that gave results that all agreed within his experimental error of 1%.

C. Thermal Conductivity

Thermal conductivity is probably the most difficult thermal property to measure accurately, and a perusal of the literature on silicon nitride shows that such is the case for this material. Data from one

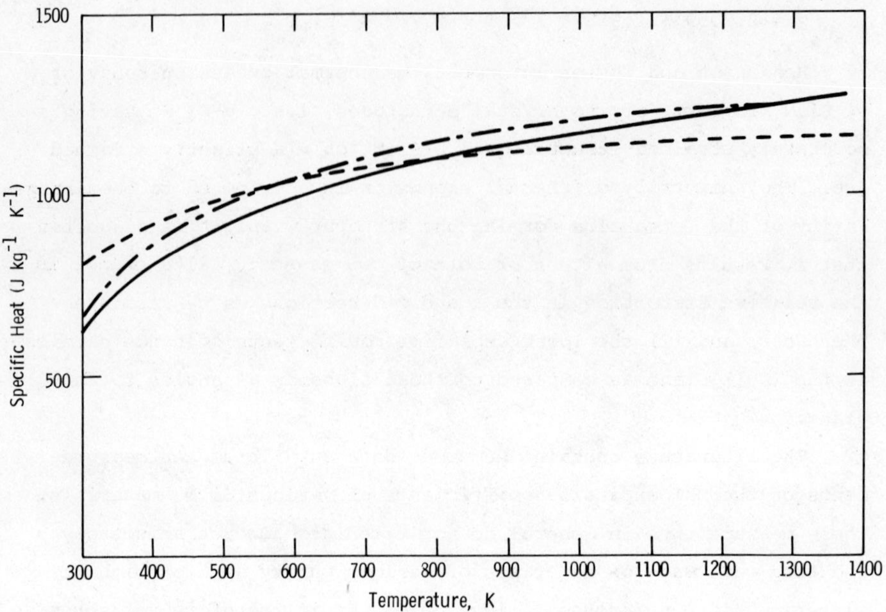

FIG. 12. Specific heat vs. temperature for silicon nitride materials; —HPSN (Miller et al., Ref. 208),----- HPSN and RBSN (George, Ref. 211), ----sintered SN (Smith and Quackenbush, Ref. 209).

of the most extensive investigations of the thermal conductivity of Si_3N_4, specifically the hot-pressed form, are shown in Fig. 13. The results shown were reported by Miller et al. [208] for specimens that were all nominally the same. The thermal conductivity not only varied from one specimen to another, but also varied considerably with the direction in the material, i.e., depending on whether the measurement was made perpendicular or parallel to the hot-pressing direction. The data of George [211], obtained on 11 Si_3N_4 materials, showed even more scatter than that shown in Fig. 13. The range of George's data, however, included that shown in the figure. It must be concluded that much more work is needed to resolve this demanding problem of generating accurate thermal conductivity data for silicon nitride materials.

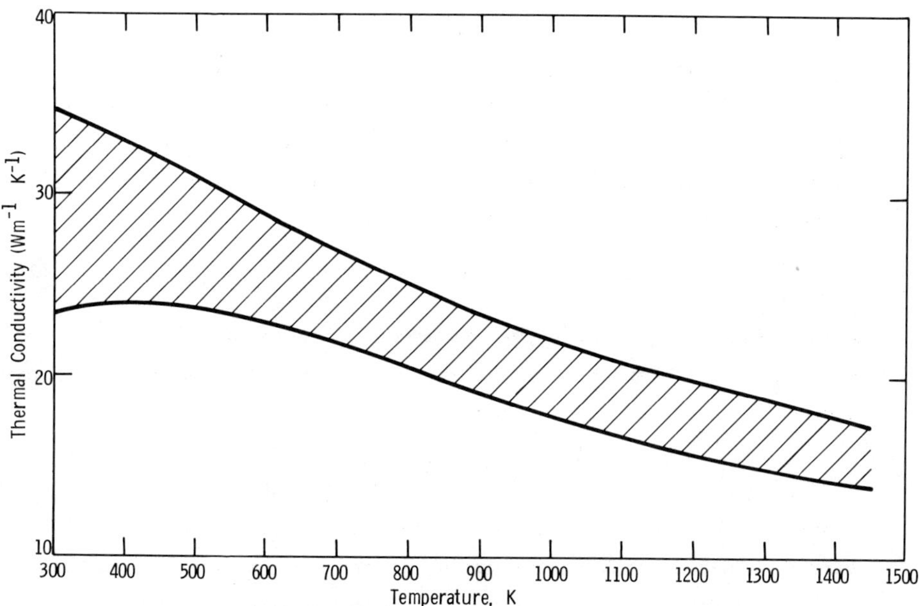

FIG. 13. Thermal conductivity of hot-pressed silicon nitride. (From Ref. 208.)

VIII. CHEMICAL BEHAVIOR

A. Oxidation

The potential uses of silicon nitride in severe high-temperature environments, e.g., gas turbines, have stimulated interest in the effects of oxidation on material properties. In addition to causing loss of material, oxidation may degrade other properties such as mechanical strength.

Chemical aspects of the oxidation of silicon nitride materials are addressed in a recent review by Singhal [212]. That review discussed the thermodynamics of the process as well as kinetic studies on the oxidation of CVD, powder, reaction-bonded, and hot-pressed Si_3N_4 materials.

Singhal [212] notes that the oxidation of Si_3N_4 at high temperatures can occur by two mechanisms. These correspond to "passive" and "active" oxidation mechanisms analogous to those proposed by Wagner [49] for the oxidation of silicon. Passive oxidation is the case where a protective SiO_2 layer is formed on the Si_3N_4 surface as represented by the following equation:

$$Si_3N_4(s) + 3O_2(g) = 3SiO_2(s) + 2N_2(g) \quad (24)$$

A specimen undergoing passive oxidation increases in weight. At sufficiently low oxygen partial pressures, active oxidation can occur according to the following equation:

$$2Si_3N_4(s) + 3O_2(g) = 6SiO(g) + 4N_2(g) \quad (25)$$

A specimen undergoing active oxidation continuously decreases in weight. In a sufficiently reducing atmosphere, surface silica can be removed from Si_3N_4 by active oxidation as follows:

$$Si_3N_4(s) + 3SiO_2(s) = 6SiO(g) + 2N_2(g) \quad (26)$$

According to Singhal, the oxygen partial pressures corresponding to the transition from active to passive oxidation range from 10^{-4} Pa at 1000°C to 10 Pa at 1600°C. All studies to date on the kinetics of oxidation of Si_3N_4 have been done in the regime of greatest practical interest, i.e., that in which passive oxidation occurs.

The few data that exist on the kinetics of oxidation of nominally pure Si_3N_4 prepared by CVD have been discussed by Singhal [212]. The oxidation of Si_3N_4 in general follows parabolic kinetics, i.e., the rate is controlled by diffusion through a SiO_2 layer. In addition, Singhal notes that, on a weight gain basis, oxidation rates are lower for pure materials than for impure ones, e.g., hot-pressed silicon nitride containing secondary phases. Surface oxidation of thin films of Si_3N_4 was studied by Raider et al. [213] who characterized the products by ESCA. Gaseous products of the oxidation of Si_3N_4 at 1000-1400°C were studied by Lin [214] using a mass spectrometric technique. He reported that the oxidation process yields

SiO_2(s) and NO(g) with the latter subsequently decomposing into N(g) and O(g).

The kinetics of oxidation of Si_3N_4 powder has been studied by a number of investigators [215-218]. All of the studies concluded that the kinetics of oxidation is parabolic, at least in the latter stages. Horton's [215] investigation was done on powder heated in dry oxygen and dry air at temperatures from 1065 to 1340°C and he obtained activation energy values respectively for the two environments of 255 and 285 kJ mol^{-1}. Goursat et al. [216,217] investigated the oxidation of Si_3N_4 powders at temperatures from 1050 to 1300°C and at oxygen partial pressures from 2 to 22 kPa and concluded that the corrosion rate was independent of O_2 pressure except at the beginning of the reaction. Goursat et al. obtained activation energy values of 293 and 146 kJ mol^{-1} for the initial and later stages of the process respectively. The former value was ascribed to an interface reaction and the latter to diffusion of oxygen in SiO_2. Mitomo and Sharp [218] investigated the oxidation of relatively pure powders of α- and β-Si_3N_4 in oxygen at temperatures to 1400°C, and concluded that the oxidation resistance of α-Si_3N_4 was less than that of β-Si_3N_4, and gave respective activation energies of 259 and 285 kJ mol^{-1} for oxidation. The activation energy value for diffusion of 145 kJ mol^{-1} obtained by Goursat et al. [216,217] is about half that found by the other workers [215,218]. Goursat et al., however, were the only workers who carefully considered powder geometry (particle size, change in interfacial area with oxidation, etc.), and their activation energy for diffusion of oxygen in SiO_2 agreed with values obtained from other measurements on the diffusion of oxygen in SiO_2. Further work may be needed to resolve this apparent disagreement among the several investigators.

The oxidation of reaction-bonded silicon nitride (RBSN) has been investigated with regard to phenomena accompanying the process and the effect of oxidation on mechanical properties. Kruse et al. [219] reported results of a thermogravimetric study in the temperature range 700-1600°C. The results were interpreted with the aid of x-ray diffraction and scanning electron microscopy, and a mechanism was suggested for the bubble formation that occurred in the oxide product layers. Lombaard and Meyer [220] investigated internal oxidation in RBSN by the MeV ^{4}He backscattering technique and concluded that an appreciable amount of oxygen diffused throughout the material. Inomata and Mitomo [221] reported that oxidation is controlled by diffusion through a cristobalite product layer, and that impregnation of RBSN with Si reduces the oxidation rate. Warburton et al. [222] reported on the oxidation of RBSN in damp air from 700 to 1100°C and gave some data on weight loss of oxidized RBSN in vacuum at 1050 to 1200°C. Barlier and Torre [223] concluded that a gaseous boundary layer formed during oxidation of RBSN in still air influences measured oxidation kinetics.

Already mentioned in a previous section was the work of Evans and Davidge [134] on the effects of oxidation on the strength of RBSN. This topic was further discussed by Davidge et al. [224]. Sharp [225] investigated by high-voltage transmission electron microscopy the internal structure of RBSN after oxidation at 1000 and 1400°C. He reported the formation of amorphous SiO_2 layers on internal pores and discussed the effect of these oxidation processes on the strength of the material. Jayatilaka and Leake [226] gave data on the effects of oxidation on the strength, elastic properties, and fracture toughness of RBSN. Already mentioned earlier was the work of Thuemmler et al. [201] which demonstrated that creep deformation observed in RBSN in oxidizing environments was largely due to the existence of oxide phases formed in such environments.

The interest in hot-pressed silicon nitride (HPSN) as a high-temperature structural material has stimulated a number of investigations of its oxidation behavior. The majority of the work to date has emphasized HPSN fabricated with MgO as a sintering aid.

Singhal [227-229] has investigated in some detail the oxidation of HPSN fabricated with MgO. His first paper [227] dealt with the corrosion of that material under static conditions at 1000-1400°C and in a pressurized gas turbine test passage at 1100°C. In a subsequent publication [228] he reported that the oxidation rate was only slightly increased by the presence of water vapor in the temperature range 1200-1400°C. A later paper [229] by Singhal treats in detail the oxidation from 1000 to 1400°C of HPSN formed with 1 w/o MgO. Oxidation was found to follow parabolic kinetics with an apparent activation energy of 375 kJ mol^{-1}, and the oxide film was found to comprise $MgSiO_3$ in which various impurity elements such as Ca, Fe, and Al tended to segregate. It was suggested by Singhal that the rate-controlling mechanism was outward diffusion of Mg ions and impurity cations through the oxide film.

Additional observations on the oxidation of the MgO material were reported by Kiehle et al. [230] and by Tripp and Graham [231]. The latter reported that the oxide scales were multiphase comprising mainly α-cristobalite and enstatite ($MgSiO_3$). Tripp and Graham also found no dependence of the oxidation rate on oxygen pressure at 1400°C.

Recent work providing additional insight into the oxidation of HPSN fabricated with MgO has been conducted by Cubicciotti et al. [232,233]. They used a volumetric technique to determine oxygen uptake over the temperature range 1248 to 1458°C and concluded that oxidation followed a parabolic law with an activation energy of 440 kJ mol^{-1}. A significant finding of these investigations, however, was that the oxide scale is not protective, i.e., that the rate of oxidation is not controlled by the rate of diffusion through the oxide scale. Electron microprobe analysis of the magnesium profile in subsurface layers indicated instead that the rate of diffusion of magnesium to the Si_3N_4 surface is the rate-limiting step in oxidation of this material.

Additional aspects of the oxidation of the MgO material have been investigated by Tighe [234] and Lange [235]. The former gives results of a study by transmission electron microscopy of changes that occur in HPSN from stresses and oxidation at high temperatures. Lange [235] related oxidation behavior to composition in the system Si_3N_4-SiO_2-MgO and noted that oxidation resistance at 1375°C was optimum for compositions near the Si_2N_4-Si_2N_2O tie line.

Although work is believed to be in progress on the oxidation of HPSN formed with additives other than MgO, the literature to date contains little information on that subject. Hampton and Graham [236] have given some results on the oxidation of HPSN materials prepared with zirconium compounds, and conclude that their oxidation resistance is better than that of the MgO material.

B. Corrosion in Other Environments

The behavior of HPSN in combustion gases such as are found in turbine environments has been the subject of several investigations. Already mentioned was the work of Singhal [227] on the corrosion of HPSN in such media. Rowcliffe and Huber [237] investigated stress corrosion of HPSN in combustion gases at 900 and 950°C and concluded that Na, Mg, and V additions do not accelerate material failure, but reported that failure in the gas environment was faster than in air. Schlichting [238] found that pyrolitic silicon nitride was more resistant than the sintered material to corrosion in hot air and CH_3 combustion products containing various salts. In another paper [239] he presented additional data on the oxidation of Si_3N_4 and its corrosion by H_2S and by Na and V compounds. Brooks and Meadowcroft [240] discuss the corrosion of HPSN and RBSN in a fuel-oil-fired environment containing Na and V (∿50 ppm of each) at 900, 1100, and 1400°C. The corrosion resistance of HPSN was found to be greater than that of RBSN at the two lower temperatures, and it was shown that corrosion involved the formation of molten Na silicates.

The effect of Na on the oxidation of RBSN was reported in two papers by Mayer and Riley [241,242]. They found that both Na_2SO_4 and NaCl vapors accelerate corrosion at 800 and 1100°C by reacting to form liquid Na silicate on the RBSN surface. They also found that, although Na_2CO_3 additions enhance initial oxidation, the enhancement terminates when specific sodium silicate compositions form.

The corrosion of HPSN and RBSN at 1000°C by molten salts comprising Na_2SO_4, NaCl, and mixtures of the two was studied by Tressler et al. [243] who concluded that, in addition to the corrosive etching that occurred, such treatment could degrade mechanical properties. Levy and Falco [244] reported that the degradation observed in RBSN after thermal cycling to 1093°C in Na_2CO_3 and Na_2SO_4 resulted from reaction between cristobalite and Na_2O to form stabilized tridymite.

Several other papers have discussed various corrosoin phenomena. Lange [245] indicated that surface pitting occurs from the reaction of Fe with HPSN fabricated with MgO, but not with HPSN fabricated with Y_2O_3. Gulden and Metcalfe [246] observed stress corrosion effects in RBSN tested in water-saturated air. Kopylova and Nazarchuk [247] gave results on the reactions of Si_3N_4 powder with acids and hydroxides.

IX. OPTICAL AND ELECTRICAL PROPERTIES

A. Optical Crystallography

Data on the crystal optics of silicon nitride were reported by Forgeng and Decker [24] and the subject was investigated in detail by Clancy [93]. Clancy's data were taken on fine transparent single and twinned crystals grown in a high-temperature, high-pressure furnace. His optical data were obtained by the immersion method using a polarizing microscope. Forgeng and Decker treated their crystals with 20% sodium hydroxide solution, which makes their values less reliable because of leaching and surface etching. Clancy's data are as follows:

α-Si_3N_4: $\omega = 2.03 \pm 0.01$ $\varepsilon = 2.02 \pm 0.01$
birefringence = (−)0.01

β-Si_3N_4: $\omega = 2.02 \pm 0.01$ $\varepsilon = 2.04 \pm 0.01$
birefringence = (+)0.02

Since both α- and β-Si_3N_4 are uniaxial crystals, ε is the refractive index in the c crystallographic direction and ω is the refractive index in the direction perpendicular to c. Clancy's paper also includes considerable information on the morphology of vapor-grown crystals.

In the case of amorphous silicon nitride films produced by chemical vapor deposition, Milek [8] reports that, using the silane-ammonia preparation technique, the refractive index of the film increases as the silane concentration increases. The refractive index values that he gives range from 2.00 to greater than 2.10. In addition, for a given silane/ammonia ratio, the refractive index decreases with increasing deposition temperature in the 800-100°C range.

B. Electrical Properties

The widespread use of silicon nitride thin films as insulators in semiconductor devices has generated much interest in the dielectric properties of such films. Milek [8] reports the bandgap of Si_3N_4 to be 4.0 eV with its optical absorption edge extending over the wavelength range from 200 to 300 nm. Considering the latter, Si_3N_4 would be expected to be transparent to visible light, and such is the case for small crystals [93] and for polycrystalline thick layers formed by CVD [26]. By the same token, bulk polycrystalline silicon nitride should be white and opaque. In fact, commercial reaction-bonded silicon nitride is characteristically dark gray in color, and Moulson [40] suggested that the darkening may result from iron impurities in that material. He further suggests that the black color of hot-pressed silicon nitride is associated with the severely reducing conditions under which it is fabricated, i.e., in graphite dies at ~1700°C.

Additional interest in the dielectric properties of Si_3N_4 has stemmed from its potential as a radome material. Such an application requires the material to withstand rapid heating to temperatures in excess of 1000°C while retaining good transparency to microwave radiation. Properties relevant to radome performance include relative dielectric constant (ε) and loss tangent ($\tan \delta$) and their variations with temperature. As pointed out by Moulson [40], these properties are sensitive to impurities and to fabrication variables, and it is therefore not surprising that the literature relating to them presents a confused picture.

Wells [248] appears to have been the first investigator to recognize that the dielectric properties of Si_3N_4 might be suitable for radome applications, and Walton [249,250] investigated the subject further. Walton [249] gave dielectric constant data for two RBSN materials of varying porosity, and indicated that the dielectric constant is related to porosity as follows:

$$\varepsilon = \varepsilon_0^{1-p} \tag{27}$$

where ε_0 is the dielectric constant for fully dense material and p the volume fraction of pores. Dielectric constant and loss data for several RBSN materials at temperatures to $\sim$1000°C were also measured by Westphal and Sils [251].

The effects of chemistry and second-phase inclusions on the dielectric properties of RBSN and HPSN were considered by Messier and Wong [252,253] and by Perry and Moules [254]. The former [253] reported that the presence of unreacted Si at levels >0.5-0.6 w/o in RBSN increased the dielectric constant significantly and dielectric loss drastically.

Although for reasons already mentioned there has been considerable variation in dielectric property values reported for RBSN, the data for relatively pure materials containing no detrimental second phases appear to be in reasonably good agreement. For material of density 2500 kg m^{-3} in the frequency range 8-10 GHz, several independent measurements [251,253,254] have yielded room-temperature values of $\varepsilon = 5.5$. Although dielectric loss seems to be more sensitive to material parameters [252-254], values of $\tan \delta < 0.005$ have been reported [249-251,253,254]. These values are well below that required ($\tan \delta < 0.01$) for radome applications [254].

Extrapolations of room-temperature dielectric constant data on porous RBSN to zero porosity [249,254] yield a value of $\varepsilon = 9.0$ for fully dense Si_3N_4. The measurements of Perry and Moules [254] on HPSN gave a similar result indicating that values ranging from ε = 6.6 to 7.4 reported for similar material [252] in the same frequency range (8-10 GHz) may be erroneous. Thorp and Sharif [255] reported results of a detailed study of the dielectric properties of HPSN (containing 5 w/o MgO) that disagree substantially with the ones quoted above. They give values of ε = 8.0-9.6 at 10^5 Hz and

$\varepsilon = 3.2\text{-}5.6$ at 9.3 GHz. The low values given at the two frequencies are for material "washed with NaOH" with no further details given. Loss tangent values reported for HPSN at 8-10 GHz range from 0.001 [252] to 0.007 [254]. The disagreement that exists concerning the dielectric properties of HPSN undoubtedly stems in part from differences in fabrication and composition of the specimens that were used. Only one report [252] gave information on the chemical purity and second-phase inclusions in the materials investigated, and Thorp and Sharif [255] reported only that their specimens contained "glass impurities."

Thorp and Sharif [256] also discussed the electrical conductivity of the HPSN materials mentioned above. The dc conductivity values that they measured went from $10^{-10}\ \Omega^{-1}\ \text{cm}^{-1}$ at 400°C to $10^{-6}\text{-}10^{-5}\ \Omega^{-1}\ \text{cm}^{-1}$ at 1000°C. The ac data taken over the frequency range 15-5 kHz indicated that a hopping process occurs below 500°C. The material was shown to be p-type below and n-type above 900°C.

In summary, it appears that obtaining precise data on the electrical properties of Si_3N_4 will require additional measurements on fully dense material of well-defined chemical purity. A promising candidate material for such measurements would be dense, polycrystalline Si_3N_4 prepared by chemical vapor deposition.

X. APPLICATIONS

As noted in Sec. I the potential usefulness of silicon nitride in demanding engineering applications has long been recognized [5,10]. Much of the early interest in the material was generated by Parr and his coworkers at the Admiralty Materials Laboratory in the United Kingdom as evidenced by papers by Parr [257] and Parr and May [258] on the applications and properties of RBSN. Despite this interest, however, a large market for silicon nitride has been slow to develop. This circumstance results largely from the demanding applications for which the material is being considered, e.g., components subjected to the high-stress, high-temperature, corrosive environments found in heat engines. A substantial development effort has existed in this area since the early seventies, and much progress has been made; large-scale use of Si_3N_4 in engines, however, awaits further development. It is beyond the scope of this chapter to cover in detail the applications of Si_3N_4, and the following text merely gives a few examples of significant applications mentioned in the recent literature.

The use of ceramic materials including Si_3N_4 in high-temperature engineering with emphasis on gas turbines has been reviewed by Godfrey [259]. The proceedings of two conferences [260,261] on the use of ceramics in high-performance applications give an excellent overview of the status of development of Si_3N_4 materials for use in systems including bearings and heat engines. A number of the individual papers in these volumes have already been cited in this chapter. Not mentioned specifically were other papers dealing with design, failure prediction, testing, etc., which also appear in Refs. 260 and 261.

Other reviews of the applications of Si_3N_4 materials have been published by Huebner [262] and Gnesin and Osipova [263]. The former considers uses in crucibles, pipes, nozzles, and tiles, and the latter applications in turbine blades, bearings, piston rings, cutting tools, and electronic components.

The largest single program to date on the application of ceramics in turbine engines is the recently concluded one sponsored by the Advanced Research Projects Agency (ARPA) of the U.S. Department of Defense and conducted by Ford Motor Co. and Westinghouse Electric Corp. The objective of that program was to develop techniques for design with brittle materials and to demonstrate the use of components fabricated from such materials in hostile environments at temperatures up to 1370°C. Demonstrated in the program were the use of stationary and rotating components comprising RBSN and RBSN/HPSN in an automotive gas turbine (Ford) and the use of HPSN stator vanes in a large gas turbine for electrical power generation (Westinghouse).

Numerous papers resulting from the ARPA program have already been cited in this chapter, and details of progress on the program are given in a series of reports [264-281]. To give a comprehensive summary of the results of a program of this magnitude in a chapter as brief as the present one is clearly impossible. It must therefore suffice to state that the program goals were substantially achieved and that operation of RBSN and HPSN components for 200 hr at temperatures up to 1370°C in an automotove gas turbine environment was demonstrated.

A program sponsored by the U.S. Navy on the applications of ceramics including Si_3N_4 in turbine engines was the subject of a recent review [282]. Also relevant is the proceedings of a workshop on ceramics in advanced heat engines [283]. These sources are also too extensive to review in detail, and they are cited merely to indicate where information on such applications can be found.

Several articles have dealt with the use of Si_3N_4 in metalworking. Nichols [284] discussed Si_3N_4 tools for drawing wire and pipe. Gnesin et al. [285] described the preparation and properties of HPSN cutting tools, and Osipova et al. [286] grinding wheels that use crushed HPSN abrasives.

The potential of HPSN as high-temperature gas bearing material was investigated by Gielisse et al. [287]. Included in their report are results of tests evaluated in terms of coefficient of friction and mode of wear.

Already mentioned was the interest in RBSN as a potential radome material [114,248-254]. Additional information on that topic appears in a report by Harris [288] who described microwave transmission tests on small radomes and mentioned the difficulty of obtaining acceptable dielectric properties in high-density ($\sim$2600 kg m^{-3}) material.

It is apparent from even a brief survey of the literature that methods for the fabrication of Si_3N_4 materials into useful end items have received considerable attention and that development is continuing at a rapid pace. It therefore seems likely that extensive applications will result for these materials in the very near future.

ACKNOWLEDGMENTS

The authors acknowledge Dr. R. N. Katz and Dr. H. F. Priest for their generous support of this effort. Special thanks go to Mrs. Helen Flionis and Ms. Diane Mancuso for typing and correcting the manuscript.

REFERENCES

1. H. Mehner, German Patent 88999, 30 Sept. 1896.
2. A. Sinding-Larsen, U.S. Patent 928,476, 20 July 1909.
3. L. Weiss and T. Engelhardt, The nitrogen compounds of silicon, *Z. Anorg. Allg. Chem. 65*, 38-65 (1910).
4. L. Wöhler, Silicon and nitrogen, *Z. Elektr. Chem. 32*, 420-23 (1926).
5. J. F. Collins and R. W. Gerby, New refractory uses for silicon nitride reported, *J. Metals 00*, 612-615 (1955).
6. I. B. Cutler and W. J. Croft, Silicon nitride, *Powder Met. Int. 6*, 92-96, 144-146 (1974).
7. D. R. Messier and M. M. Murphy, *An Annotated Bibliography on Silicon Nitride for Structural Applications*, Metals and Ceramics Information Center, Battelle Columbus Laboratory, Columbus, Ohio, MCIC-79-41, 1979.
8. J. T. Milek, Silicon nitride for microelectronic applications--Part 1, Preparation and properties, *Handbook of Electronic Materials*, Vol. 3, IFI/Plenum New York, 1971.
9. M. Billy, Preparation and characterization of silicon nitride, *Ann. Chim.* (Paris) *4*, 795-851 (1959).
10. E. T. Turkdogan, P. M. Bills, and V. A. Tippet, Silicon Nitride: Some physicochemical properties, *J. Appl. Chem. 8*, 296-302 (1958).
11. D. Hardie and K. H. Jack, Crystal structure of silicon nitride, *Nature 180*, 332-333 (1957).
12. R. Juza and H. Hahn, The crystal structures of Zn_3N_2, Cd_3N_2, and Ge_3N_4, *Z. Anorg. Chem. 224*, 125-132 (1949).
13. S. N. Ruddlesden and P. Popper, On the crystal structures of the nitrides of silicon and germanium, *Acta Cryst. 11*, 465-468 (1958).
14. S. Wild, P. Grieveson, and K. H. Jack, The crystal structures of alpha and beta silicon and germanium nitrides. In *Special Ceramics 5* (P. Popper, ed.), British Ceramic Research Association, England, 1972, pp. 385-395.
15. O. Borgen and H. M. Seip, The crystal structure of beta-Si_3N_4, *Acta Chem. Scand. 15*, 1789 (1961).
16. K. Narita and K. Mori, Crystal structures of silicon nitride, *Bull. Chem. Soc. Japan 32*, 417-419 (1959).
17. R. Marchand, Y. Laurent, J. Lang, and M. Th. LeBihan, The structure of alpha silicon nitride, *Nitrure de Silicium Alpha, Acta Cryst. B25*, 2157-60 (1969).
18. I. Kohatsu and James W. McCauley, Re-examination of the crystal structure of α-Si_3N_4, *Mat. Res. Bull. 9*, 917-920 (1974).

19. K. Kato, Z. Inoue, K. Kijima, I. Kawada, H. Tanaka and T. Yamane Structural approach to the problem of oxygen content in alpha silicon nitride, *J. Amer. Ceram. Soc. 58*, 90-91 (1975).

20. H. F. Priest, F. C. Burns, G. L. Priest, and E. Skaar, The oxygen content of alpha silicon nitride, *J. Amer. Ceram. Soc. 56*, 395 (1973).

21. A. J. Edwards, D. P. Elias, M. W. Lindley, A. Atkinson, and A. J. Moulson, Oxygen content of reaction bonded α silicon nitride, *J. Mat. Sci 9*, 516-517 (1974).

22. D. S. Thompson and P. L. Pratt, The structure of silicon nitride, in *Science of Ceramics 3* (G. H. Stewart, ed.), Academic Press, 1967, pp. 33-51.

23. Donald R. Messier, *The Alpha/Beta Silicon Nitride Phase Transformation*, Army Materials and Mechanics Research Center, Watertown, Mass., AMMRC TR 77-15, 1977.

24. W. D. Forgeng and B. F. Decker, Nitrides of silicon, *Trans. Metl. Soc. AIME 212*, 343-348 (1958).

25. H. Suzuki, The synthesis and properties of silicon nitride, *Bull. Tokyo Inst. of Technol.*, No. 54, 163-177 (1963).

26. F. Galasso, U. Kuntz, and W. J. Croft, Pyrolytic Si_3N_4, *J. Amer. Ceram. Soc. 55*, 431 (1972).

27. P. Grieveson, K. H. Jack, and S. Wild, The crystal structures of alpha and beta silicon and germanium nitrides. In *Special Ceramics 4*, 1968, pp. 237-238.

28. S. Wild, P. Grieveson, and K. H. Jack, Thermodynamic and phase relations in the silicon-nitrogen-oxygen system, in *Metallurgical Chemistry*, Proceedings of a Symposium, 1971 (O. Kubaschewski, ed.), Her Majesty's Stationery Office, London, 339-346 (1972).

29. S. Wild, P. Grieveson, and K. H. Jack, The thermodynamics and kinetics of formation of phases in the Ge-N-O and Si-N-O systems. In *Special Ceramics 5*, 1972, pp. 271-288.

30. I. Colquhoun, S. Wild, P. Grieveson, and K. H. Jack, Thermodynamics of the silicon-nitrogen-oxygen system, *Proc. Brit. Ceram. Soc.*, No. 22, 207-227 (1973).

31. H. Feld, P. Ettmayer, and I. Petzenhauser, Oxygen stabilization of α-Si_3N_4, *Ber. Deut. Keram. Ges. 51*, 127-131 (1974).

32. W. B. Hincke and L. R. Brantley, The high temperature equilibrium between silicon nitride, silicon, and nitrogen, *J. Amer. Chem. Soc. 52*, 48-52 (1930).

33. R. D. Pehlke and J. F. Elliott, High temperature thermodynamics of the silicon, nitrogen, silicon-nitride system, *Trans. AIME 215*, 781-785 (1959).

34. *JANAF Thermochemical Tables*, 2nd ed., U.S. Govt. Printing Office, Washington, D.C., 1971.

35. K. Blegen, Equilibria and kinetics in the system Si-N and Si-N-O. In *Special Ceramics 6*, 1975, pp. 223-244.

36. S. C. Singhal, Thermodynamic analysis of the high-temperature stability of silicon nitride and silicon carbide, *Ceramur. Int. 2*, 123-130 (1976).

37. S. A. Jansson and E. A. Gulbransen, in *High Temperature Gas-Metal Reactions in Mixed Environments* (S. A. Jansson and Z. A. Foroulis, eds.), TMS-AIME, New York, 1973.

38. E. A. Ryklis, A. S. Bolgar, and V. V. Fesenko, Evaporation and thermodynamic properties of silicon nitride, *Sov. Powder Met. Metal Ceram.*, No. 1, 73-76 (1969) [translation of *Porosh. Met.*, No. 1, 92-96 (1969)].

39. N. D. Potter and G. Piper, Thermodynamic Properties of High Temperature Materials, AFOSR Contract Report, AFOSR-74-0799Tr, 1974.

40. A. J. Moulson, Reaction-bonded silicon nitride: Its formation and properties, *J. Mater. Sci. 14*, 1017-1051 (1979).

41. F. L. Riley, Nitridation and reaction bonding, in *Nitrogen Ceramics* (F. L. Riley, ed.), Noordhoff, Leyden, The Netherlands, 1977, pp. 265-288.

42. P. Popper and S. N. Ruddlesden, The preparation, properties, and structure of silicon nitride, *Trans. Brit. Ceram. Soc. 60*, 603-626 (1961).

43. D. R. Messier and P. Wong, Kinetics of nitridation of Si powder compacts, *J. Amer. Ceram. Soc. 56*, 480-485 (1973).

44. O. Glemser, K. Beltz, and P. Naumann, The silicon nitrogen system, *Z. Anorg. Allg. Chem. 291*, 51-66 (1957).

45. J. W. Evans and S. K. Chatterji, Kinetics of the oxidation and nitridation of silicon at high temperatures, *J. Phys. Chem. 62*, 1064-1067 (1958).

46. K. Mueller and H. Hollerer, Der Einfluss von Katalysorten auf die Bildungsgeschwindigkeit von Siliziumnitrid, Vortrag Sweite Konferenze fur Pulvermetallurgie, Krakau, 1967.

47. K. J. Huettinger, On the kinetics of nitridation of silicon, *High Temp. High Press. 1*, 221-230 (1969).

48. D. R. Messier, P. Wong, and A. E. Ingram, Effect of oxygen impurities on the nitridation of high-purity silicon, *J. Amer. Ceram. Soc., 56*, 171-172 (1973).

49. C. Wagner, Passivity during oxidation of silicon at elevated temperatures, *J. Appl. Phys. 29*, 1295-1297 (1958).

50. R. S. Wagner and W. C. Ellis, Vapor-liquid solid mechanism of crystal growth and its application to silicon, *Trans. AIME 233*, 1053-1064 (1965).

51. V. N. Gribkov, V. A. Silaev, B. V. Shchetanov, E. L. Umantsev, and A. S. Isaikin, Growth mechanism of silicon nitride whiskers, *Sov. Phys.-Crystallogr. 16*, 852-854 (1972) [translation of *Kristallografyia 16*, 982-985 (1971)].

52. A. Atkinson, P. J. Leatt, and A. J. Moulson, The nitriding of silicon powder compacts, *J. Mater. Sci. 7*, 482-484 (1972).

53. A. Atkinson, P. J. Leatt, and A. J. Moulson, The role of nitrogen flow into the nitriding compact in the production of reaction-sintered silicon nitride, *Proc. Brit. Ceram. Soc.*, No. 22, 253-274 (1973).

54. R. B. Guthrie and F. L. Riley, The nitridation of single-crystal silicon, *Proc. Brit. Ceram. Soc.*, No. 22, 275-280 (1973).

55. R. B. Guthrie and F. L. Riley, Effect of oxide impurities on the nitridation of high-purity silicon, *J. Mater. Sci 9*, 1363-1365 (1974).

56. A. Atkinson, P. J. Leatt, A. J. Moulson, and E. W. Roberts, A mechanism for the nitridation of silicon powder compacts, *J. Mat. Sci. 9*, 981-984 (1974).

57. A. Atkinson, A. J. Moulson, and E. W. Roberts, Nitridation of high-purity silicon. *J. Mater. Sci. 10*, 1242-1243 (1975).

58. A. Atkinson, A. J. Moulson, and E. W. Roberts, Nitridation of high-purity silicon, *J. Amer. Ceram. Soc. 59*, 285-288 (1976).

59. D. Campos-Loriz and F. L. Riley, The effect of silica on the nitridation of silicon, *J. Mater. Sci. 11*, 195-198 (1976).

60. S. M. Boyer, D. Sang, and A. J. Moulson, The effects of iron on the nitridation of silicon, in *Nitrogen Ceramics* (F. L. Riley, ed.), Noordhoff, Leyden, The Netherlands, 1977, pp. 297-304.

61. D. Campos-Loriz and F. L. Riley, Factors affecting the formation of the α- and β-phases of silicon nitride, *J. Mat. Sci. 13*, 1125-1127 (1978).

62. P. Longland and A. J. Moulson, The growth of α- and β-Si_3N_4 accompanying the nitriding of silicon powder compacts, *J. Mater. Sci. 13*, 2279-2280 (1978).

63. S. M. Boyer and A. J. Moulson, A mechanism for the nitridation of Fe-contaminated silicon, *J. Mater. Sci. 13*, 1637-1646 (1978).

64. W. M. Dawson and A. J. Moulson, The combined effects of Fe and H_2 on the kinetics of silicon nitridation, *J. Mater. Sci. 13*, 2289-2290 (1978).

65. S. S. Lin, Mass spectrometric studies of the nitridation of silicon, *J. Amer. Ceram. Soc. 58*, 271-273 (1975).

66. Y. Inomata and Y. Uemura, Nitridation kinetics of silicon powder, *Yogyo Kyokai Shi 83*, 244-248 (1975).

67. Y. Inomata, Nitridation of silicon powder, *Yogyo Kyokai Shi 83*, 497-500 (1975).

68. W. Naruse, M. Nojiri, and M. Tada, Formation conditions and properties of the two crystal phases of silicon nitride, *Nippon Kinzoku Gakkaishi 35*, 731-738 (1971).

69. M. Mitomo, Effects of oxygen partial pressure on nitridation of silicon, *J. Amer. Ceram. Soc. 58*, p. 527 (1975).

70. M. Mitomo, Effect of Fe and Al additions on nitridation of silicon, *J. Mater. Sci. 12*, 273-276 (1977).

71. A. de S. Jayatilaka and J. A. Leake, The role of impurities in the formation of reaction-bonded silicon nitride. In *Nitrogen Ceramics* (F. L. Riley, ed.), Noordhoff, Leyden, The Netherlands, 1977, p. 289-295.

72. D. R. Messier and R. L. Riley, The α/β silicon nitride phase transformation. In *Nitrogen Ceramics*, (F. L. Riley, ed.), Noordhoff, Leyden, The Netherlands, 1977, pp. 141-149.

73. D. R. Messier, F. L. Riley, and R. J. Brook, The α/β silicon nitride phase transformation, *J. Mater. Sci. 13*, 1199-1205 (1978).

74. R. F. Coe, R. J. Lumby, and M. F. Pawson, Some properties and applications of hot-pressed silicon nitride, *Special Ceramics 5*, 361-376 (1972).

75. R. J. Lumby and R. F. Coe, The influence of some process variables on the mechanical properties of hot-pressed silicon nitride *Proc. Br. Ceram. Soc. (15)*, 91-101 (1970).

76. S. Wild, P. Grieveson, K. H. Jack, and M. J. Latimer, The role of magnesia in hot-pressed silicon nitride, *Special Ceramics 5*, 377-384 (1972).

77. A. G. Evans and J. V. Sharp, Microstructural studies on silicon nitride, *J. Mater. Sci. 6*, 1292-1302 (1971).

78. R. J. Brook, T. G. Carruthers, L. J. Bowen, and R. J. Weston, Mass transport in the hot pressing of α-silicon nitride, in *Nitrogen Ceramics*, (F. L. Riley, ed.), Noordhoff, Leyden, The Netherlands, 383-392 (1977).

79. H. Knoch and G. Ziegler, Influence of MgO content and temperature on transformation kinetics, grain structure and mechanical properties of hot-pressed silicon nitride, *Science of Ceramics 9*, 494-501 (1977).

80. L. J. Bowen, R. J. Weston, T. G. Carruthers, and R. J. Brook, Hot-pressing and the α-β phase transformation in silicon nitride, *J. Mater. Sci. 13*, 341-350 (1978).

81. R. J. Weston and T. G. Carruthers, Kinetics of hot-pressing of alpha-silicon nitride powder with additives, *Proc. Br. Ceram. Soc. (22)*, 197-206 (1973).

82. Y. Inomata, Stability relation in the system β-silicon nitride-α-silicon nitride-silicon oxynitride (Si_2N_2O) and their structural change by heating above 1600°, *Yogyo Kyokai Shi 82*, 522-526 (1974).

83. P. Drew and M. H. Lewis, The microstructures of silicon nitride ceramics during hot-pressing transformations, *J. Mater. Sci. 9*, 261-269 (1974).

84. P. Drew and M. H. Lewis, The microstructures of silicon nitride/alumina ceramics, *J. Mater. Sci. 9*, 1833-1838 (1974).

85. K. Nuttall and D. P. Thompson, Observations on the microstructures of hot-pressed silicon nitride, *J. Mater. Sci. 9*, 850-853 (1974).

86. R. Kossowsky, Wetting of silicon nitride by alkaline-doped $MgSiO_3$, *J. Mater. Sci. 9*, 2025-2033 (1974).

87. K. H. Jack, New developments in silicon nitride ceramics, in *Ceramics for High Performance Applications*, (J. J. Burke, A. E. Gorum, and R. N. Katz, ed.), Brook Hill Publishing, Chestnut Hill, Mass., 1975, pp. 265-286.

88. G. R. Terwilliger and F. F. Lange, Pressureless sintering of Si_3N_4, *J. Mater. Sci. 10*, 1169-1174 (1975).

89. C. Greskovich and S. Prochazka, Observations on the α-β Si_3N_4 transformation, *J. Amer. Ceram. Soc. 60*, 471-472 (1977).

90. C. M. B. Henderson and D. Taylor, Thermal expansion of nitrides and oxynitride of silicon in relation to their structures, *Trans. J. Brit. Ceram. Soc. 74*, 49-53 (1975).

91. M. J. Buerger, *Phase Transformations in Solids*, (R. Smoluchowski, J. E. Mayer, and W. A. Weyl, eds.), John Wiley, New York, 1951, p. 183.

92. A. Verma and P. Krishna, *Polymorphism and Polytypism in Crystals*, John Wiley, New York, 1966.

93. W. P. Clancy, A limited crystallographic and optical characterization of alpha and beta silicon nitride, *Microscope 22*, 279-315 (1974).

94. W. D. Glaeser, Alpha phase silicon nitride production: Some aspects of nitriding kinetics and powder morphology, in *Ceramics for High Performance Applications-II* (J. J. Burke, E. M. Lenoe, and R. N. Katz, eds.), Brook Hill Publishing, Chestnut Hill, Mass., 1978, pp. 549-557.

95. A. Kato, Y. Ono, S. Kawazoe, and I. Mochida, Finely divided silicon nitride by vapor phase reaction between silicon tetrachloride and ammonia, *Yogyo Kyokai Shi 80*, 114-120 (1972) (in Japanese).

96. K. S. Mazdiyasni and C. M. Cooke, Synthesis, characterization, and consolidation of Si_3N_4 obtained from ammonolysis of $SiCl_4$, *J. Amer. Ceram. Soc. 56*, 628-633 (1973).

97. M. Mitomo, H. Tanaka, and J. Tanaka, The synthesis of α-Si_3N_4, *Yogyo Kyokai Shi 82*, 144-145 (1974).

98. V. F. Funke and G. V. Samsonov, Preparation and certain properties of silicon nitride, *Zh. Absch. Khim. 28*, 267-272 (1958) (in Russian).

99. K. Komeya and H. Inoue, Synthesis of the α-form of silicon nitride from silica, *J. Mater. Sci. 10*, 1243-1246 (1975).

100. A. Hendry and K. H. Jack, The preparation of silicon nitride from silica. In *Special Ceramics 6*, (P. Popper, ed.), British Ceramic Research Association, England, 1975, pp. 199-208.

101. S. Motoi and S. Hidaka, Synthesis of silicon nitride from silica, *Denki Kagaku Ayobi Kogyo Butsuri Kogaku 43*, 33-38 (1975), (in Japanese).

102. I. S. Brokhin and V. F. Funke, The preparation of silicon nitride and the investigation of some of its properties, *Ogneupory 22*, 562-566 (1957) (in Russian).

103. N. L. Parr, Silicon nitride, a new ceramic for high temperature engineering and other applications, *Research* (London) *13*, 261-269 (1960).

104. N. L. Parr, G. F. Martin, and E. R. W. May, Preparation, Microstructure, and mechanical properties of silicon nitride, *Special Ceramics* (P. Popper, ed.), Academic Press, New York, 1960, pp. 102-135.

105. N. L. Parr, R. Sands, P. L. Pratt, E. R. W. May, C. R. Shakespeare and D. S. Thompson, Structural aspects of silicon nitride, *Powder Met. 8*, 152-163 (1961).

106. N. L. Parr, Silicon nitride, *The Engineer*, *222*, 18-19 (1966).

107. N. L. Parr and E. R. W. May, The technology and engineering applications of reaction-bonded silicon nitride, *Proc. Br. Ceram. Soc. (7)*, 81-98 (1967).

108. A. Rabeneau, Silicon nitride, a ceramic material for high temperatures, *Ber. Dtsch. Keram. Ges. 40*, 6-12 (1963) (in German).

109. D. J. Godfrey, The fabrication and properties of silicon nitride ceramics and their relevance to aerospace applications, *J. Br. Interplanetary Soc. 22*, 353-368 (1969).

110. R. L. Brown, D. J. Godfrey, M. W. Lindley, and E. R. W. May, Advances in the technology of silicon nitride ceramics, *Special Ceramics 5*, (P. Popper, ed.), British Ceramic Research Association, Manchester, England, 345-360 (1972).

111. R. L. Brown, The flame sprayed deposition of silicon powder for the production of silicon nitride ceramics, *Metal Construction Brit. Weld. J. 1*, 317-321 (1969).

112. T. M. Valentine and D. I. Malkin, Foamed silicon nitride, *Proc. Brit. Ceram. Soc. (22)*, 281-289 (1973).

113. D. I. Matkin, I. E. Denton, T. M. Valentine, and P. Warrington, Fabrication of silicon nitride by ceramic/plastic technology, *Proc. Brit. Ceram. Soc. (22)*, 291-304 (1973).

114. J. N. Harris, Slip-cast reaction sintered silicon nitride for radome applications, *Proc. 12th Symp. on Electromagnetic Windows* (J. N. Harris, ed.), Georgia Institute of Technology, Atlanta, Georgia, 72-75 (1974).

115. A. Ezis, The fabrication and properties of a slip-cast silicon nitride, in *Ceramics for High Performance Applications*, (J. J. ing, Chestnut Hill, Mass., 1975, p. 207-222.

116. C. F. Johnson and T. G. Mohr, Injection molding 2.7 g/cc silicon nitride turbine rotor blade rings utilizing automatic control, *Ceramics for High Performance Applications-II*, (J. J. Burke, E. M. Lenoe, and R. N. Katz, eds.), Brook Hill Publishing, Chestnut Hill, Mass., 1978, p. 194-206.

117. W. Engel, E. Lange, and N. Mueller, Injection molded and duo density silicon nitride, in *Ceramics for High Performance* Brook Hill Publishing, Chestnut Hill, Mass, p. 527-538 (1978).

118. A. Atkinson and A. J. Moulson, Some important variables affecting the course of the reaction between silicon powder and nitrogen, *Science of Ceramics 8*, British Ceramic Society, Stoke-on-Trent, England, 111-121 (1976).

119. S.-S. Lin, Comparative studies of metal additives on the nitridation of silicon, *J. Amer. Ceram. Soc. 60*, 78-81 (1977).

120. S.-S. Lin, Comparison of iron-alloy additives on nitridation of silicon, *J. Amer. Ceram. Soc. 61*, 95-96 (1978).

121. I. Amato, D. Martorana, and M. Rossi, Nitriding of silicon powder compacts, *Powder Met. 18*, 339-348 (1975).

122. B. J. Dalgleish and P. L. Pratt, The microstructure of reaction-bonded silicon nitride, *Proc. Brit. Ceram. Soc. (22)*, 323-326 (1973).

123. B. F. Jones and M. W. Lindley, Strength, density, nitrogen weight gain relationships for reaction sintered silicon nitride, *J. Mater. Sci. 10*, 967-972 (1975).

124. B. F. Jones and M. W. Lindley, Strength/density relationships in partially nitrided silicon compacts - Their use in reaction sintered silicon nitride research and technology, *Science of Ceramics 8*, British Ceramic Society, Stoke-on-Trent, 123-132 (1976).

125. B. F. Jones and M. W. Lindley, Strength, density, and nitrogen weight gain relationships for reaction sintered silicon nitride prepared from fine silicon powders, *Powder Met. Int. 8*, 32-34 (1976).

126. B. F. Jones and M. W. Lindley, Additional observations on the strength/nitrided density relationship for reaction sintered silicon nitride, *J. Mater. Sci. 11*, 191-193 (1976).

127. D. P. Elias, B. F. Jones, and M. W. Lindley, The formation of α- and β-phases in reaction sintered silicon nitride and their influence on strength, *Powder Met. Int. 8*, 162-165 (1976).

128. D. P. Elias and M. W. Lindley, Reaction sintered silicon nitride, Part 1: The influence of oxygen and water vapour contamination on strength and composition, *J. Mater. Sci. 11*, 1278-1287 (1976).

129. B. F. Jones and M. W. Lindley, Reaction sintered silicon nitride, Part 2: The influence of nitrogen gas flow on strength and strength/density relationships, *J. Mater. Sci. 11*, 1288-1295 (1976).

130. B. F. Jones and M. W. Lindley, The influence of hydrogen in the nitriding gas on the strength of reaction sintered silicon nitride, *J. Mater. Sci. 11*, 1969-1971 (1976).

131. B. F. Jones, K. C. Pitman, and M. W. Lindley, The development of strength in reaction sintered silicon nitride, *J. Mat. Sci. 12*, 563-576 (1977).

132. B. F. Jones, K. C. Pitman and M. W. Lindley, A structural model explaining the development of strength in reaction sintered silicon nitride, in *Nitrogen Ceramics* (F. L. Riley, ed.), Noordhoff, Leyden, The Netherlands, 1977, pp. 561-567.

133. M. W. Lindley, K. C. Pitman, and B. F. Jones, Reaction Sintered Si_3N_4: Development of mechanical properties relative to microstructure and nitriding environment, *Fracture Mechanics of Ceramics*, Vol. 4, *Crack Growth and Microstructure* (R. C. Brandt, D. P. H. Hasselman, and F. F. Lange, eds.), Plenum Press, New York, 1978, pp. 921-932.

134. A. G. Evans and R. W. Davidge, The strength and oxidation of reaction-sintered silicon nitride, *J. Mater. Sci. 5*, 324-325 (1970).

135. D. J. Godfrey and M. W. Lindley, The strength of reaction-bonded silicon nitride ceramics, *Proc. Brit. Ceram. Soc.*, No. 22, 229-252 (1973).

136. D. R. Messier and P. Wong, Kinetics of formation and mechanical properties of reaction sintered Si_3N_4, in *Ceramics for High Performance Applications* (J. J. Burke, A. E. Gorum, and R. N. Katz, Brook Hill Publishing, Chestnut Hill, Mass., 1975, pp. 181-194.

137. D. J. Godfrey, The effects of impurities, additions, and surface preparation on the strength of silicon nitride, *Proc. Brit. Ceram. Soc.*, No. 25, 325-337 (1975).

138. D. J. Godfrey and K. C. Pitman, Some mechanical properties of silicon nitride ceramics: Strength, hardness, and environmental effects, in *Ceramics for High Performance Applications* (J. J. Burke, A. E. Gorum, and R. N. Katz, eds.), Brook Hill Publishing, Chestnut Hill, Mass., 1975, pp. 425-444.

139. J. A. Mangels, Effects of H_2-N_2 nitriding atmosphere on the properties of reaction-sintered Si_3N_4, *J. Amer. Ceram. Soc. 58*, 354-355 (1975).

140. J. A. Mangels, Strength-density-nitriding cycle relationships for reaction-sintered Si_3N_4, *Nitrogen Ceramics* (F. L. Riley, ed.), Noordhoff, Leyden, The Netherlands, 1977, p. 569-574.

141. J. Heinrich, Einfluss der Ausgangskorngrösse von Silicium auf das Gefüge und die Mechanischen Eigenschaften von reaktion-gesintertem Silicium-nitrid, *Ber. Deut. Keram. Ges. 55*, 238-241 (1978).

142. P. Wong and D. R. Messier, Procedure for fabrication of Si_3N_4 by rate-controlled reaction sintering, *Amer. Ceram. Soc. Bull. 57*, 525-526 (1978).

143. J. A. Mangels, Development of injection molded reaction bonded Si_3N_4, in *Ceramics for High Performance Applications*, Vol. II (J. J. Burke, E. M. Lenoe, and R. N. Katz, eds.), Brook Hill Publishing, Chestnut Hill, Mass., 1978, pp. 113-130.

144. Y. Inomata, Oxidation resistant Si-impregnated surface layer on reaction sintered silicon nitride articles, *Yogyo Kyokai Shi 83*, 1-3 (1975).

145. G. Leimer and E. Gugel, Infiltrated composites based on silicon nitride, *Z. Metallk. 66*, 570-576 (1975).

146. K. S. Mazdiyasni, R. West, and L. R. David, Characterization of organosilicon-infiltrated porous reaction-sintered Si_3N_4, *J. Amer. Ceram. Soc. 61*, 504-508 (1978).

147. G. G. Deely, J. M. Herbert, and N. C. Moore, Dense silicon nitride, *Powder Met.*, No. 8, 145-151 (1961).

148. A. F. Fickel and H. Kessel, Technology and applications of dense silicon nitride, *Glas. Email. Keram. Tech. 25*, 193-196 (1974).

149. E. Gugel, A. F. Fickel, and H. Kessel, Developments in the production of hot-pressed silicon nitride, *Powder Met. Int. 6*, 136-140 (1974).

150. A. Hivert, Performance of silicon nitride radone tips, Proc. 3rd Internat. Conf. Electromagnetic Windows, Délégation Ministérielle pour l'Armement, Paris, 1976.

151. R. J. Bratton, C. A. Andersson, and F. F. Lange, Hot-pressed Si_3N_4 developments, in *Ceramics for High Performance Applications*, Vol. II (J. J. Burke, E. M. Lenoe, and R. N. Katz, eds.), Brook Hill Publishing Co., Chestnut Hill, Mass., 1978, pp. 805-825.

152. Y. Oyama and O. Kamigaito, Sintered silicon nitride-magnesium oxide system, *Yogyo Kyokai Shi 81*, 290-293 (1973).

153. I. Colquhoun, D. P. Thompson, W. I. Wilson, P. Grieveson, and K. H. Jack, The determination of surface silica and its effect on the hot-pressing behavior of alpha-silicon nitride powder, *Proc. Brit. Ceram. Soc.*, No. 22, 181-195 (1973).

154. G. R. Terwilliger and F. F. Lange, Hot-pressing behavior of Si_3N_4, *J. Amer. Ceram. Soc. 57*, 25-29 (1974).

155. I. I. Osipova and D. A. Pogorelova, Recrystallization of silicon nitride during hot pressing, *Sov. Powder Met. Metal Ceram. 14*, 1015-1017 (1975) [translation of *Porosh. Met. 14*, 74-77 (1975)].

156. L. J. Bowen, R. J. Weston, T. G. Carruthers, and R. J. Brook, Mechanisms of densification during the pressure sintering of alpha-silicon nitride, *Ceramurg. Int. 2*, 173-176 (1976).

157. J. L. Iskoe, F. F. Lange, and E. S. Diaz, Effect of selected impurities on the high temperature mechanical properties of hot-pressed silicon nitride, *J. Mater. Sci. 11*, 908-912 (1976).

158. H. Knoch and G. Ziegler, Influences of powder compositions and sintering temperature on transformation kinetics, structure, and mechanical properties of hot pressed silicon nitride, *Ber. Deut. Keram. Ges. 55* 242-245 (1978).

159. J. C. Uy, R. M. Williams, and M. U. Goodyear, Study of the processing and bond strength of hot pressed Si_3N_4, in *Ceramics for High Performance Applications*, Vol. II (J. J. Burke, E. M. Lenoe, and R. N. Katz, eds.), Brook Hill Publishing, Chestnut Hill, Mass., 1978, pp. 131-149.

160. L. J. Bowen and T. G. Carruthers, Development of mechanical strength in hot-pressed silicon nitride, *J. Mat. Sci. 13*, 684-687 (1978).

161. G. E. Gazza, Hot pressed Si_3N_4, *J. Amer. Ceram. Soc. 56*, 662 (1973).

162. G. E. Gazza, Effect of yttria additions on hot-pressed Si_3N_4, *Amer. Ceram. Soc. Bull. 54*, 778-781.

163. A. Tsuge, K. Nishida, and M. Komatsu, High-temperature strength of hot-pressed Si_3N_4 containing Y_2O_3, *J. Amer. Ceram. Soc. 58*, 323-326 (1975).

164. G. Q. Weaver and J. W. Lucek, Optimization of hot-pressed Si_3N_4-Y_2O_3 materials, *Amer. Ceram. Soc. Bull. 57*, 1131-1134 (1978).

165. L. J. Bowen, T. G. Carruthers, and R. J. Brook, Hot-pressing of Si_3N_4 with Y_2O_3 and Li_2O additives, *J. Amer. Ceram. Soc. 61*, 335-339 (1978).

166. A. Tsuge and K. Nishida, High strength hot pressed Si_3N_4 with concurrent Y_2O_3 and Al_2O_3 additions, *Amer. Ceram. Soc. Bull. 57*, 424-426, 431 (1978).

167. R. R. Wills, Silicon yttrium oxynitrides, *J. Amer. Ceram. Soc. 57*, 459 (1974).

168. A. Tsuge, H. Kudo, and K. Komeya, Reaction of Si_3N_4 and Y_2O_3 in hot-pressing, *J. Amer. Ceram. Soc. 57*, 269-270 (1974).

169. A. W. J. M. Rae, D. P. Thompson, N. J. Pipkin and K. H. Jack, The structure of yttrium silicon oxynitride and its role in the hot-pressing of silicon nitride with yttria additions, in *Special Ceramics*, Vol. 6 (P. Popper, ed.), British Ceramic Research Association, Manchester, England, 1975, pp. 347-360.

170. G. E. Gazza, H. Knoch, and G. D. Quinn, Hot-pressed Si_3N_4 with improved thermal stability, *Amer. Ceram. Soc. Bull. 57*, 1059-1060 (1978).

171. R. W. Rice, W. J. McDonough, and P. F. Becher, Zirconia additions to hot pressed silicon nitride, *Rep. NRL Progr.*, 18-20 (December 1974).

172. R. W. Rice and W. J. McDonough, Hot-pressed Si_3N_4 with Zr based additives, *J. Amer. Ceram. Soc. 58*, 264 (1975).

173. I. C. Huseby and G. Petzow, Influence of various densifying additives on hot-pressed Si_3N_4, *Powder Met. Int. 6*, 17-19 (1974).

174. J. E. Weston, P. L. Pratt, and B. C. H. Steele, Crystallization of grain boundary phases in hot-pressed silicon nitride materials, Part 1--Preparation and characterization of materials, *J. Mat. Sci. 13*, 2137-2146 (1978).

175. J. E. Weston and P. L. Pratt, Crystallization of grain boundary phases in hot-pressed silicon nitride materials, Part 2--Mechanical properties of materials, *J. Mater. Sci. 13*, 2147-2156 (1978).

176. G. R. Terwilliger, Properties of sintered Si_3N_4, *J. Amer. Ceram. Soc. 57*, 48-49 (1974).

177. G. E. Gazza, Sintered silicon nitride, in *Ceramics for High Performance Applications*, Vol. II (J. J. Burke, E. M. Lenoe, and R. N. Katz, eds.), Brook Hill Publishing, Chestnut Hill, Mass., 1978, pp. 1001-1010.

178. V. M. Sleptsov, O. D. Shcherbina, A. N. Pilyankevich, and G. S. Oleinik, Strength and microstructure of materials based on silicon nitride, *Poroshk. Metall.*, No. 8, 83-87 (1976).

179. M. Mitomo, C. Oshima, and M. Tsutsumi, Microstructural change during gas pressure sintering of silicon nitride, *Yogyo Kyokai Shi* (in Japanese) *84*, 356-360 (1976).

180. M. Mitomo, M. Tsutsumi, E. Bannai, and T. Tanaka, Sintering of Si_3N_4, *Amer. Ceram. Soc. Bull. 55*, 313 (1976).

181. H. F. Priest, G. L. Priest, and G. E. Gazza, Sintering of Si_3N_4 under high nitrogen pressure, *J. Amer. Ceram. Soc. 60*, 81 (1977).

182. T. I. Mah, K. S. Mazdiyasni, and R. Ruh, The role of cerium orthosilicate in the densification of Si_3N_4, *J. Amer. Ceram. Soc. 62*, 12-16 (1979).

183. C. Greskovich, S. Prochazka, and J. H. Roslowski, The sintering behavior of covalently-bonded materials, in *Nitrogen Ceramics* (F. L. Riley, ed.), Noordhoff, Leyden, The Netherlands, 1977, p. 351-357.

184. C. Greskovich and C. O'Clair, Effect of impurities on sintering Si_3N_4 containing MgO or Y_2O_3 additive, *Amer. Ceram. Soc. Bull. 57*, 1055-1056 (1978).

185. H. Masaki and O. Kamigaito, Pressureless sintering of silicon nitride with additions of MgO, Al_2O_3, and/or Spinel, *Yogyo Kyokai Shi 84*, 508-512 (1976).

186. M. Mitomo, Sintering of Si_3N_4 with Al_2O_3 and Y_2O_3, *Yogyo Kyokai Shi 85*, 408-412 (1977).

187. I. Oda, M. Kameno, and N. Yamamoto, Pressureless sintered silicon nitride, in *Nitrogen Ceramics* (F. L. Riley, ed.), Noordhoff, Leyden, The Netherlands, 1977, pp. 359-365.

188. F. S. Galasso, R. D. Veltri, and W. J. Croft, Chemically vapor deposited Si_3N_4, *Amer. Ceram. Soc. Bull. 57*, 453-454 (1978).

189. K. Kijima, N. Setaka, and H. Tanaka, Preparation of silicon nitride single crystals by chemical vapor deposition, *J. Cryst. Growth 24/25*, 183-187 (1974).

190. K. Niihara and T. Hirai, Chemical vapour-deposited silicon nitride, Part I--Preparation and some properties, *J. Mat. Sci. 11*, 593-603 (1976).

191. K. Niihara and T. Hirai, Chemical vapour-deposited silicon nitride, Part II--Density and formation mechanism, *J. Mat. Sci. 11*, 604-611 (1976).

192. K. Niihara and T. Hiari, Chemical vapour-deposited silicon nitride, Part III--Structural features, *J. Mat. Sci. 12*, 1233-1242 (1977).

193. K. Niihara and T. Hirai, Chemical vapour-deposited silicon nitride, Part IV--Hardness characteristics, *J. Mat. Sci. 12*, 1243-1252 (1977).

194. W. F. Knippenberg and G. Verspui, Influence of impurities on the growth of silicon carbide crystals, Silicon Carbide, Proc. 2nd Int. Conf., 1968, pp. 33-44.

195. K. Kijima, N. Setaka, M. Ishii, and H. Tanaka, Preparation of Si_3N_4 by vapour-phase reaction, *J. Amer. Ceram. Soc. 56*, 346 (1973).

196. J. W. Edington, D. J. Rowcliffe, and J. L. Henshall, Powder metallurgical review, 8: The mechanical properties of silicon nitride and silicon carbide, Part I--Materials and strength, *Powder Met. Int. 7*, 82-96 (1975).

197. J. W. Edington, D. J. Rowcliffe, and J. L. Henshall, Powder metallurgical review, 8: The mechanical properties of silicon nitride and silicon carbide, Part II--Engineering Properties, *Powder Met. Int. 7*, 136-147 (1975).

198. F. F. Lange, Strong high-temperature ceramics, in *Annual Review of Materials Science* (R. A. Huggins, ed.), Annual Reviews, Inc., Palo Alto, Calif., 1974, pp. 365-390.

199. J. A. Mangels, Development of a creep resistant reaction sintered Si_3N_4, in *Ceramics for High Performance Applications* (J. J. Burke, A. E. Gorum, and R. N. Katz, eds.), Brook Hill Publishing, Chestnut Hill, Mass., 195-206.

200. M. E. Washburn and H. R. Baumgartner, High temperature properties of reaction-bonded silicon nitride, in *Ceramics for High Performance Applications* (J. J. Burke, A. E. Gorum, and R. N. Katz, eds.), Brook Hill Publishing, Chestnut Hill, Mass., 1975, pp. 479-492.

201. F. Thuemmler, F. Porz, G. Grathwohl, and W. Engel, Creep and oxidation of reaction sintered Si_3N_4, in *Science of Ceramics 8*, British Ceramic Society, Stoke-on-Trent, England, 1976, pp. 133-144.

202. J. A. Mangels, High temperature, time dependent physical property characterization of reaction sintered Si_3N_4, in *Nitrogen Ceramics* (F. L. Riley, ed.), Noordhoff, Leyden, The Netherlands, 1977, pp. 589-594.

203. R. Kossowsky, Creep fatigue of Si_3N_4 as related to microstructure, in *Ceramics for High Performance Applications* (J. J. Burke, A. E. Gorum, and R. N. Katz, eds.), Brook Hill Publishing Co., Chestnut Hill, Mass., 1975, 347-372.

204. R. Kossowsky, D. G. Miller, and E. S. Diaz, Tensile and creep strengths of hot-pressed Si_3N_4, *J. Mat. Sci. 10*, 983-997 (1975).

205. A. Tsuge, K. Nishida, and M. Komatsu, High-temperature strength of hot-pressed Si_3N_4 containing Y_2O_3, *J. Amer. Ceram. Soc. 58*, 323-326 (1975).

206. F. F. Lange, Dense Si_3N_4: Interrelation between phase equilibria, microstructure, and mechanical properties, in *Nitrogen Ceramics* (F. L. Riley, ed.), Noordhoff, Leyden, The Netherlands, 1977, 491-509.

207. G. D. Quinn, *Characterization of Turbine Ceramics After Long Term Environmental Exposure*, Army Materials and Mechanics Research Center, Watertown, Mass., AMMR CTR 80-15, April 1980.

208. D. G. Miller, C. A. Andersson, S. C. Singhal, F. F. Lange, E. S. Diaz, and R. Kossowsky, *Brittle Materials Design, High Temperature Gas Turbine*, Army Materials and Mechanics Research Center, Watertown, Mass., AMMRC CTR 76-32, Vol. IV, 1976.

209. J. T. Smith and C. L. Quackenbush, GTE Si_3N_4-based ceramics, presented at Highway Vehicles Systems Contractor's Coordination Meeting, Dearborn, Mich., 24-26 April, 1979.

210. D. C. Larsen and G. C. Walther, *Property Screening and Evaluation of Ceramic Turbine Engine Materials*, IIT Research Institute, Chicago, IITRI-D6114-ITR-35, 1978.

211. W. George, Thermal property measurements on silicon nitride and silicon carbide ceramics between 290 and 700 K, *Proc. Brit. Ceram. Soc.*, No. 22, 147-167 (1973).

212. S. C. Singhal, Oxidation of silicon nitride and related materials, in *Nitrogen Ceramics* (F. L. Riley, ed.), Noordhoff, Leyden, The Netherlands, 1977, pp. 607-626.

213. S. I. Raider, R. Flitsch, J. A. Aboaf, and W. A. Pliskin, Surface oxidation of silicon nitride films, *J. Electrochem. Soc. 123*, 560-565 (1976).

214. S.-S. Lin, Mass spectrometric analysis of vapors in oxidation of Si_3N_4 compacts, *J. Amer. Ceram. Soc. 58*, 160 (1975).

215. R. M. Horton, Oxidation kinetics of powdered silicon nitride, *J. Amer. Ceram. Soc. 52*, 121-124 (1969).

216. P. Goursat, P. Lortholary, D. Tetrard, and M. Billy, Silicon nitride and oxynitride stability in oxygen at high temperatures, Proc. 7th Int. Symp. Reactivity of Solids, Bristol, U.K., 17-21 July 1972 (J. S. Anderson, M. W. Roberts, and P. S. Stone, eds.), Chapman and Hall, London, 1972, pp. 315-326.

217. D. Tetrard, P. Lortholary, P. Goursat, and M. Billy, Research on the nitrides of silicon, IV: Kinetics of oxidation of Si_3N_4 powder, *Rev. Int. Hautes Temp. Refract. 10*, 153-159 (1973).

218. M. Mitomo and J. H. Sharp, Oxidation of α- and β-silicon nitride, *Yogyo Kyokai Shi 84*, 33-36 (1976).

219. B. D. Kruse, G. Willmann, and H. Hausner, Oxidation of reaction-sintered silicon nitride, *Ber. Deut. Keram. Ges. 53*, 349-351 (1976).

220. J. M. Lombaard and O. Meyer, Depth profiling of oxygen in oxidized ceramic silicon nitride (Si_3N_4) with the backscattering technique, *Nucl. Instrum. Methods 145*, 347-351 (1977).

221. Y. Inomata and M. Mitomo, Oxidation of silicon nitride and oxidation inhibition, *Muki Zaishitsu Kenkysusho Kenkyu Hokokusho 13*, 44-49 (1977).

222. J. B. Warburton, J. E. Antill, and R. W. M. Hawes, Oxidation of thin sheet reaction-sintered silicon nitride, *J. Amer. Ceram. Soc. 61*, 67-72 (1978).

223. P. Barlier and J. P. Torre, On the influence of a gaseous boundary layer on the oxidation of reaction-bonded silicon nitride at 1400°C, *J. Mat. Sci. 14*, 235-237 (1979).

224. R. W. Davidge, A. G. Evans, D. Gilling, and P. R. Wilyman, Oxidation of reaction-sintered silicon nitride and effects on strength, in *Special Ceramics*, Vol. 5 (P. Popper, ed.), British Ceramic Research Association, Manchester, England, 1972, pp. 329-344.

225. J. V. Sharp, Electron microscopy of oxidized silicon nitride, *J. Mat. Sci. 8*, 1755-1758 (1973).

226. A. deS. Jayatilaka and J. A. Leake, The influence of progressive oxidation on the fracture toughness and strength of reaction-bonded silicon nitride at elevated temperatures in air, *Proc. Brit. Ceram. Soc.*, No. 25, 311-323 (1975).

227. S. C. Singhal, Oxidation and corrosion-erosion behavior of Si_3N_4 and SiC, in *Ceramics for High Performance Applications* (J. J. Burke, A. E. Gorum, and R. N. Katz, eds.), Brook Hill Publishing, Chestnut Hill, Mass., 1974, pp. 533-548.

228. S. C. Singhal, Effect of water vapor on the oxidation of hot-pressed silicon nitride and silicon carbide, *J. Amer. Ceram. Soc. 59*, 81-82 (1976).

229. S. C. Singhal, Thermodynamics and kinetics of oxidation of hot-pressed silicon nitride, *J. Mat. Sci. 11*, 500-509 (1976).

230. A. J. Kiehle, L. K. Heung, P. J. Gielisse, and T. J. Rockett, Oxidation behavior of hot-pressed Si_3N_4, *J. Amer. Ceram. Soc. 58*, 17-20 (1975).

231. W. C. Tripp and H. C. Graham, Oxidation of Si_3N_4 in the range 1300-1500°C, *J. Amer. Ceram. Soc. 59*, 399-403 (1976).

232. D. Cubicciotti, K. H. Lau, and R. L. Jones, The rate controlling process in the oxidation of hot-pressed silicon nitride, *J. Electrochem. Soc. 124*, 1955-1956 (1977).

233. D. Cubicciotti and K. H. Lau, Kinetics of oxidation of hot-pressed silicon nitride containing magnesia, *J. Amer. Ceram. Soc. 61*, 512-517 (1978).

234. N. J. Tighe, Microstructural aspects of deformation and oxidation of magnesia-doped silicon nitride, in *Nitrogen Ceramics* (F. L. Riley, ed.), Noordhoff, Leyden, The Netherlands, 1977, pp. 441-448.

235. F. F. Lange, Phase relations in the system Si_3N_4-SiO_2-MgO and their interrelation with strength and oxidation, *J. Amer. Ceram. Soc. 61*, 53-56 (1978).

236. A. F. Hampton and H. C. Graham, Oxidation of hot-pressed silicon nitride with zirconium-based additives, *Oxid. Met. 10*, 239-253 (1976).

237. D. J. Rowcliffe and P. A. Huber, Hot gas stress corrosion of silicon nitride and silicon carbide, *Proc. Brit. Ceram. Soc.*, No. 25, 239-252 (1975).

238. J. Schlichting, Hot corrosion behavior of SiC and Si_3N_4 in combustion gases, *Werkst. Korros. 26*, 753-758 (1975).

239. J. Schlichting, Oxidation and hot corrosion behavior of Si_3N_4 and Sialon, in *Nitrogen Ceramics* (F. L. Riley, ed.), Noorhdoff, Leyden, The Netherlands, 1977, pp. 627-634.

240. S. Brooks and D. B. Meadowcroft, The corrosion of silicon based ceramics in a residual fuel oil fired environment, *Proc. Brit. Ceram. Soc.*, No. 26, 237-250 (1978).

241. M. I. Mayer and F. L. Riley, Sodium ion assisted oxidation of reaction-bonded silicon nitride, *Proc. Brit. Ceram. Soc.*, No. 26, 251-264 (1978).

242. M. I. Mayer and F. L. Riley, Sodium assisted oxidation of reaction-bonded silicon nitride, *J. Mat. Sci. 13*, 1319-1328 (1978).

243. R. E. Tressler, M. D. Meiser, and T. Yonushonis, Molten salt corrosion of SiC and Si_3N_4 ceramics, *J. Amer. Ceram. Soc. 59*, 278-279 (1976).

244. M. Levy and J. J. Falco, Hot corrosion of reaction-bonded Si_3N_4, *Amer. Ceram. Soc. Bull. 57*, 457-458 (1978).

245. F. F. Lange, Reaction of iron with Si_3N_4 materials to produce surface pitting, *J. Amer. Ceram. Soc. 61*, 270-271 (1978).

246. M. E. Gulden and A. G. Metcalfe, Stress corrosion of silicon nitride, *J. Amer. Ceram. Soc. 59*, 391-396 (1976).

247. V. P. Kopylova and T. N. Nazarchuk, Chemical stability of silicon nitride and oxynitride powders, *Sov. Powder Met. Metal Ceram. 14*, 812-816 (1976) [translation of *Porosh. Met. 14*, 38-43 (1975)].

248. W. M. Wells, Silicon nitride as a high temperature radome material, Lawrence Radiation Laboratory, Livermore, Calif., UCRL 7795, 1964.

249. J. D. Walton, Jr., Reaction sintered silicon nitride as a hypersonic radome material, Proc. 11th Symp. Electromagnetic Windows (N. E. Poulos and J. D. Walton, Jr., eds.), Georgia Institute of Technology, Atlanta, 1972, pp. 103-106.

250. J. D. Walton, Jr., Reaction sintered silicon nitride for high temperature radome applications, *Bull. Amer. Ceram. Soc., 53*, 255-258 (1974).

251. W. B. Westphal and A. Sils, *Dielectric Constant and Loss Data*, Air Force Materials Laboratory, Wright-Patterson Air Force Base, Ohio, AFML-TR-72-39, 1972.

252. D. R. Messier and P. Wong, Silicon nitride: A promising material for radome applications, Proc. 12th Symp. Electromagnetic Windows (J. N. Harris, ed.), Georgia Institute of Technology, Atlanta, 1974, pp. 62-66.

253. D. R. Messier and P. Wong, Effect of processing conditions on microwave dielectric properties of reaction-sintered silicon nitride, Proc. 13th Symp. Electromagnetic Windows (H. L. Bassett and J. M. Newton, eds.), Georgia Institute of Technology, Atlanta, 1976, pp. 3-8.

254. G. S. Perry and T. R. Moules, Microwave dielectric properties of silicon nitrides, Proc. 12th Symp. Electromagnetic Windows (J. N. Harris, ed.), Georgia Institute of Technology, Atlanta, 1974, pp. 67-71.

255. J. S. Thorp, and R. I. Sharif, Dielectric properties of some hot-pressed nitrogen ceramics, *J. Mat. Sci. 12*, 2274-80 (1977).

256. J. S. Thorp, and R. I. Sharif, Electrical conductivity in hot-pressed nitrogen ceramics, *J. Mat. Sci. 11*, 1494-1500 (1976).

257. N. L. Parr, Silicon Nitride, A new ceramic for high temperature engineering and other applications, *Research* (London) *13*, 261-269 (1960).

258. N. L. Parr and E. R. W. May, The technology and engineering applications of reaction-bonded silicon nitride, *Proc. Brit. Ceram. Soc.*, No. 7, 81-98 (1967).

259. D. J. Godfrey, Ceramics for high-temperature engineering?, *Proc. Brit. Ceram. Soc.*, No. 22, 1-25 (1973).

260. J. J. Burke, A. E. Gorum, and R. N. Katz, eds., *Ceramics for High Performance Applications*, Brook Hill Publishing, Chestnut Hill, Mass., 1975.

261. J. J. Burke, E. M. Lenoe, and R. N. Katz, eds., *Ceramics for High Performance Applications*, Vol. 2, Brook Hill Publishing, Chestnut Hill, Mass., 1978.

262. K. Huebner, Silicon Nitride, A ceramic material with outstanding temperature-, shock-, and corrosion-resistance, *Chem-Ztg. 95*, 931-934 (1971).

263. G. G. Gnesin and I. I. Osipova, Wear-resistant engineering materials based on silicon nitride, *Russ. Eng. J. 54*, 33-36 (1974) [translation of *Vestn. Mashinostr.*, No. 12, 32-34 (1974)].

264. A. F. McLean, E. A. Fisher, and D. E. Harrison, Brittle Materials Design, High Temperature Gas Turbine, Interim Report, 1 July 1971-31 December 1971, Army Materials and Mechanics Research Center, Watertown, Mass., AMMRC CTR 72-3, 1972.

265. A. F. McLean, E. A. Fisher, and R. J. Bratton, Brittle Materials Design, High Temperature Gas Turbine, Interim Report, 1 January 1972-30 June 1972, Army Materials and Mechanics Research Center, Watertown, Mass., AMMRC CTR 72-19, 1972.

266. A. F. McLean, E. A. Fisher, and R. J. Bratton, Brittle Materials Design High Temperature Gas Turbine, Interim Report, 1 July 1972-31 December 1972, Army Materials and Mechanics Research Center, Watertown, Mass., AMMRC CTR 73-9, 1973.

267. A. F. McLean, E. A. Fisher, and R. J. Bratton, Brittle Materials Design, High Temperature Gas Turbine, Interim Report, 1 January 1973-30 June 1973, Army Materials and Mechanics Research Center, Watertown, Mass., AMMRC CTR 73-32, 1973.

268. A. F. McLean, E. A. Fisher, and R. J. Bratton, Brittle Materials Design High Temperature Gas Turbine, Interim Report No. 5, 1 July 1973-31 December 1973, Army Materials and Mechanics Research Center, Watertown, Mass., AMMRC CTR 74-26, 1974.

269. A. F. McLean, E. A. Fisher, and R. J. Bratton, Brittle Materials Design High Temperature Gas Turbine, Interim Report No. 6, 1 January 1974-30 June 1974, Army Materials and Mechanics Research Center, Watertown, Mass., AMMRC CTR 74-59, 1974.

270. A. F. McLean, E. A. Fisher, R. J. Bratton, and D. G. Miller, Brittle Materials Design, High Temperature Gas Turbine, Interim Report No. 7, 1 July 1974-31 December 1974, Army Materials and Mechanics Research Center, Watertown, Mass., AMMRC CTR 75-8 (1975).

271. A. F. McLean, E. A. Fisher, R. J. Bratton, and D. G. Miller, Brittle Materials Design, High Temperature Gas Turbine, Interim Report No. 8, 1 January 1975-30 June 1975, Army Materials and Mechanics Research Center, Watertown, Mass., AMMRC CTR 75-28, 1975.

272. A. F. McLean, E. A. Fisher, R. J. Bratton, and D. G. Miller, Brittle Materials Design, High Temperature Gas Turbine, Interim Report No. 9, 1 July 1975-31 December 1975, Army Materials and Research Center, Watertown, Mass., AMMRC CTR 76-12, 1976.

273. A. F. McLean and R. R. Baker, Brittle Materials Design, High Temperature Gas Turbine, Interim Report No. 10, 1 January 1976-30 June 1976, Army Materials and Mechanics Research Center, Watertown, Mass., AMMRC CTR 76-31, 1976.

274. A. F. McLean and E. A. Fisher, Brittle Materials Design, High Temperature Gas Turbine, Interim Report No. 11, 1 July 1976-31 December 1976, Army Materials and Mechanics Research Center, Watertown, Mass., AMMRC CTR 77-20, 1977.

275. A. F. McLean and R. R. Baker, Brittle Materials Design, High Temperature Gas Turbine, Vol. 1, Ceramic Component Fabrication and Demonstration, Interim Report No. 12, 1 January 1977-30 September 1977, Army Materials and Mechanics Reserach Center, Watertown, Mass., AMMRC TR 78-14-Vol. 1, 1978.

276. A. F. McLean and R. R. Baker, Brittle Materials Design, High Temperature Gas Turbine, Vol. 2, Ceramic Turbine Rotor Technology, Army Materials and Mechanics Research Center, Watertown, Mass., AMMRC TR 78-14-Vol. 2, 1978.

277. A. F. McLean and R. R. Baker, Brittle Materials Design, High Temperature Gas Turbine, Interim Report No. 13, 1 October 1977-31 March 1978, Army Materials and Mechanics Research Center, Watertown, Mass., AMMRC TR 79-11, 1979.

278. R. J. Bratton and D. G. Miller, Brittle Materials Design, High Temperature Gas Turbine, Final Report, Westinghouse Electric Corp., Army Materials and Mechanics Research Center, Watertown, Mass., AMMRC CTR 76-32-Vol. I, 1976.

279. D. G. Miller and C. R. Booher, Jr., Brittle Materials Design, High Temperature Gas Turbine Stator Vane Development and Static Rig Tests, Army Materials and Mechanics Research Center, Watertown, Mass., AMMRC CTR 76-32-Vol. II, 1976.

280. D. G. Miller and W. Van Buren, Brittle Materials Design, High Temperature Gas Turbine Rotor Blade Development and Turbine Modification, Army Materials and Mechanics Research Center, Watertown, Mass., AMMRC CTR 76-32-Vol. III, 1976.

281. D. G. Miller, C. A. Andersson, S. C. Singhal, F. F. Lange, E. S. Diaz, and R. Kossowsky, Brittle Materials Design, High Temperature Gas Turbine Material Technology, Army Materials and Mechanics Research Center, Watertown, Mass., AMMRC CTR 76-32-Vol. IV, 1976.

282. Proc. 1977 DARPA/NAVSEA Ceram. Gas Turb. Demonstr. Eng. Progr. Rev. (J. W. Fairbanks and R. W. Rice, eds.), Battelle Columbus Laboratories, Columbus, Ohio, MCIC-78-36, 1978.

283. Proc. Workshop Ceram. Advan. Heat Eng., Orlando, Fla., 24-26 January 1977, Energy Research and Development Administration, Washington, D.C., CONF-770110, 1977.

284. D. C. Nichols, Silicon nitride as a working material for noncutting forming tools, *Draht-Welt 63*, 434-436 (1977).

285. G. G. Gnesin, I. I. Osipova, M. M. Mai, V. N. Paderno, and V. P. Yaroshenko, Preparation and Properties of Tool Materials Based on Silicon Nitride, Grundlagen, Herstellung Eigenschaften Sinter-Verbundenwerkst [Abh. Int. Pulvermetall. Tag. Vorabdrucke], 6th, 2, Paper No. 45, 1977.

286. I. I. Osipova, A. A. Adamovskii, S. S. Krizhanovskii, and V. N. Paderno, An abrassive material based on silicon nitride for grinding unhardened steels, *Sov. Powder Met. Metal Ceram. 16*, 348-351 (1977) [translation of *Porosh. Met.*, No. 5, 34-37 (1977)].

287. P. J. Gielisse, M. P. Wilson, V. N. Borase, L. K. Heung, and A. J. Kiehle, Ceramic Materials for High Temperature Gas Bearings, Rhode Island University, Kingston, Office of Naval Research, URI-9804-4046, 1972.

288. J. N. Harris, Investigation of Reaction Sintered Silicon Nitride as a Radome Material, Georgia Institute of Technology, Atlanta, Naval Air Systems Command, GIT-A-1585-F, 1975.

AUTHOR INDEX

Numbers in parentheses are reference numbers and indicate that an author's work is referred to although the name may not be cited in the text. Numbers in italics give the page on which the complete reference is cited.

SUBJECT INDEX